ANTHRACITE LADS
A True Story Of The Fabled Molly Maguires

By
WILLIAM H. BURKE

Printed in Canada
ISBN No. 1-883658-47-0
1 2 3 4 5 6 7 8 9 0

ACKNOWLEDGMENTS

Many contributed to this work, including authors with whose conclusions I disagree. They were working to unravel a finely spun conspiracy concealed by jury verdicts and hysterical public opinion. They performed much of the labor required to blast the truth away from the gob. I am indebted to them.

My special thanks to the Honorable John P. Lavelle for the historical research and legal analysis contained in his work, *The Hard Coal Docket*, and for providing me with trial transcripts, newspaper articles, and kind encouragement. Howard Crown, author of *A Guide To The Molly Maguires* and *A Molly Maguire On Trial*, not only provided me copies of his trial transcripts in the *Shenandoah Herald* and clippings from other newspapers, but he spent considerable time making available his insights gleaned from his lifetime of study of the Mollies. I never met Patrick Campbell, grand-nephew of Alex Campbell and author of *A Molly Maguire Story* and *Who Killed Franklin Gowen?*, but I'm indebted to him for his ground-breaking research and analysis. My chapters about the death of Gowen were inspired by him.

My thanks to Ray Burke and Bob Rosini for reading my first drafts and making corrections and suggestions for improvement. Dr. David Frew of the Erie County Historical Society has long been a font of encouragement and advice. His corrections are in the text. Of course, the mistakes are mine alone. Finally, I wish to express special thanks to my wife Mary for her love, support and patience with my mumblings about Molly Maguire.

William H. Burke

Introduction To

ANTHRACITE LADS
A True Story Of The Fabled Molly Maguires

The Author's Search For The Truth

"The Molly Maguires!" That dreaded name was spoken only in whispers in the hard coal region when the author was a boy. Caution was advisable. The descendants of the men who were hanged, or of those who had caused them to be hanged, might well be within earshot and take fresh offense.

The Mollies were reported by the newspapers of their time to be a widespread Irish criminal conspiracy infiltrated by a lone Pinkerton detective, and brought to justice at the end of a rope by Reading Railroad President Franklin B. Gowen. Many historical works confirm this version of early organized crime in America's hard coal region. It was popularized by Sir Arthur Conan Doyle's Sherlock Holmes novel, *The Valley of Fear*, and by the critically acclaimed 1970 film, *The Molly Maguires*.

But the author believes that the newspapers and the writers of history have perpetrated or accepted a deliberately contrived fiction – a myth. In this work he seeks to reveal the truth, and to tell that truth in a way that will generate a widespread understanding of the real significance of the Molly Maguire saga in America's history.

His great-grandfather and grandfather were miners who, with their families, lived through the Molly era in the house on Tioga Street depicted in the book. He was raised there. He experienced the very real pressures that have led to so much violence in the anthracite coal region. He has always known that no fable was needed to explain that violence.

Upon reading all the books published about the Molly Maguires, he wasn't satisfied that they made sense. An undercover Pinkerton detective was supposed to have joined the Mollies and gained their confidence and brought them all to justice. But the men weren't convicted of the crimes that had caused the newspapers to claim there was a gang called the Mollies in the first place. After the men were hanged, those crimes continued. Furthermore, many of the crimes for which alleged Mollies were hanged could easily have been prevented by the detective, but he didn't do it. In fact, he encouraged and helped plan the crimes, by his own admission. The books of the more recent historians struggle to resolve these discrepancies, but they fail to do it.

The author left his coal-region roots to attend Harvard Law School and to practice law. It occurred to him that, perhaps, uncovering the truth about the Mollies required a legal analysis of their trial transcripts. A transcript of the first Molly trial, of one Dan Dougherty, was found in the pages of the old *Pottsville Miners' Journal*. Reading it, the author was astonished to discover that every historian who had written about it had misstated and misunderstood it. They misstate the testimony about the murder weapon and a bullet matched to it. This causes them to presume that whether Dougherty was guilty or innocent was a close question. But it was not. As the transcript shows, Dougherty was clearly innocent. The judge practically instructed the jury to find him innocent. The historians were led to overlook the true significance of the case. It was the ethnically motivated prosecution of a clearly innocent man who was shot at for being a Molly even after he was acquitted on overwhelming evidence.

The author learned that in the early 1980's the handwritten transcripts of the trials in Carbon County had been discovered in the basement of the courthouse. They became available on film. When the author studied the testimony of the vaunted Pinkerton detective, James McParlan, in the Alex Campbell case, he got another shock. Most historians had praised the detective's testimony as clear, organized and unshakeable. The author shows it to be rambling, confusing, self-contradictory and even laughably unbelievable. Campbell's conviction on the strength of such testimony was a gross miscarriage of justice. What the testimony does show is that the Pinkertons planned, encouraged and permitted the murder, and not Campbell.

When the Molly Maguire papers of the Reading Railroad became available in the Hagley Museum, Wilmington, Delaware, the author obtained copies. Only one historian, W.G. Broehl, had reviewed the field operative's reports and expense accounts. He had concluded that they support McParlan's story. But a comparison of the detective's activity reports with his expense accounts does not support the agent. Most notably, the report for July 4, 1875, claims that he spent all day in Girardville with the accused King of the Mollies, John Kehoe, getting the King's help to hire killers. But the agent's expense account shows he spent that day buying drinks at Big Mine Run, the domain of Kehoe's enemy and archrival. This and other entries in the Pinkertons' own records prove that McParlan was not a credible witness and, in fact, a liar and a drunk.

The detective reports also show that, contrary to the assumptions of historians, McParlan was discovered to be a detective by the Shamokin men long before he dramatically appeared in open court to testify against the Mollies. The lads lied to him that they were going to wreck the passenger train to Pottsville. As expected, he notified the police who came swarming to catch them in the act. They walked some men down the tracks whistling, with their hands in their pockets, to observe the police in ambush. They did this on several nights, as documented in the reports of the Pinkerton operative Captain Linden.

Further, the author's legal background caused him to recognize that while Allan Pinkerton's *The Molly Maguires And The Detectives* is " . . . a biased, self-serving version . . . ," of a " . . . contrived industrial fable . . . " as stated by the editor who republished it in 1973, it is nevertheless a contemporaneous statement by one of the principal actors in the cases.

It is the equivalent of a declaration by a party to the cases. Therefore, we must consider seriously the many damaging admissions which it contains, such as:

❖ The Pinkerton operative known as McKenna was the actual gangleader of the thugs in Shenandoah who were called Mollies.

❖ This supposedly clear-eyed detective was a petty thief and a consumer of heroic quantities of alcohol.

❖ McKenna was actually looking forward to testifying against the Mollies and revealing himself as a detective. He had never exacted a promise from his employers that he would not be required to testify, as he claimed at the trials.

❖ The Shenandoah gang of Mollies who committed many of the crimes were boon-drinking companions of Coal & Iron Police Captain Robert P. Linden and were confident he would overlook their crimes.

Finally, the author's contacts with coal-region historians made him aware of their ground-breaking research. They show how all Irishmen were excluded from the jury panels and that the detective known as McKenna was not a lone operator who hadn't the ability to warn of murders about to be committed. He was accompanied to the coal-region by two of his brothers. They lurk as enigmatic, shadowy figures in the historical records. But their existence may now be noted as bearing upon whether McKenna and the Pinkertons ever had any intention of preventing any murders.

Placing each of these discoveries (or rediscoveries) in its proper place refashions the fabric of the story as a whole. The truth comes out. The fabled Molly Maguire criminal conspiracy was, in fact, Gowen and his Pinkertons aided by vigilantes and some Irish thugs cultivated in varying degrees of ignorance.

AUTHOR'S APOLOGIA –
For Telling The Truth This Way

Possessing the truth requires an answer to the question of how best to tell it. The Molly Maguire myth had hanged twenty men, destroyed countless more lives, distracted the nation from horrific casualty rates and child labor in the mining industry, and defamed an entire class of immigrants. I believe that such a deadly fiction should be destroyed in the most public way possible. It is not a task for a tome of dry history. Legal concepts such as "preponderance of the evidence" and "probability" are most useful in judging the proof of disputed facts. And the truth is exciting! But a proper writing form is required to bring these tools to the task.

To make the story readable for the public I chose to tell it as a first-person eye-witness account. This form is usually reserved for fiction, but with the proper disclosures it can be usefully employed to encourage an understanding of the truth.

While the narrator and his family are not found in the historical records, people much like them did exist and were noted. Although the narrator and his kin interact with historical figures it is always in a way which the records indicate that someone did so interact with them. For example, the records show that someone did convince the undercover detective that a train was going to be derailed, and caused him to telegraph Captain Linden about it, and observed as the squad of police waited in the bushes to apprehend the wreckers. The narrator and his family are the kind of people who, in all probability, had done that. Giving a name and a face to such persons and describing what were probably their motives does not make a work untrue. All that is required is a caveat warning the reader to be on guard to exercise independent judgment.

And this leads me to a further caveat – about the last two chapters which depict the shooting of Railroad President Franklin B. Gowen. All the history books with a single exception, state that Gowen locked himself in a room in the Wormley Hotel in Washington, D.C.

and blew his own brains out. I disagree. The forensic evidence discloses that to be impossible. It is physically impossible for a man to shoot himself from above and behind the right ear with a Smith & Wesson revolver at such a distance as to leave no powder burns. This conclusion is supported by the opinions of two modern forensics experts.

The truth is that someone shot Gowen, and before the coroner arrived the owner of the hotel moved the body, tore up the carpet in the room, and ripped off the wallpaper. Later, for reasons known only to Gowen's family and to Captain Linden (whom they hired to investigate) the detective disregarded the overwhelming contradictory evidence and issued a finding that Gowen had shot himself.

The facts of the case are ably stated by Patrick Campbell in his work, *Who Killed Franklin Gowen?*

Presented herein is a version of the incident which I admit is only one hypothesis of what might have happened. However, this version does take into account all the known facts of the case and attempts to reconcile them. This is what judges instruct juries to do when attempting to discern what really happened from the statements of multiple witnesses. I am confident this version is closer to the truth than the fiction perpetrated by the Pinkertons and preserved in the history tomes – that Gowen committed suicide. It is not logical to assume that the first-person narrative form cannot or should not be used to tell the truth.

Finally, this little work is not intended to exhaust the subject of the Molly Maguires. There were too many men hanged, too many trials, and too many lives destroyed to do full justice to them all in a single book. To permit the casual reader to appreciate the fundamentals of the story without undue confusion, I name only the principal actors in the drama and deal with only those events which best illustrate the essential truths which have been obscured by jury verdicts and hysterical public opinion. My hope is to foment a dialogue which will encourage others to continue the search for the truth and write about what they find.

Editor's Note On
The Manuscript

This manuscript came to us from a dead man. It was delivered by one of the nation's most prestigious law firms, along with sufficient funds to defray the costs of publication. We would have rejected the offer had the material been unsuitable, but we are convinced that the manuscript, as well as the manner of its presentation, may be part of the solution to mysteries which have baffled historians for a hundred and thirty years.

The work seems to be the first person narrative of Matthew McWilliams, Esq., who in the 1870's, had been a coal miner and union organizer with close ties to Irishmen whom the newspapers of the time characterized as Molly Maguires. They were reported to be an early example of organized crime in America. But Attorney McWilliams uses Pinkerton Detective Agency reports and his legal analysis of trial transcripts to reveal a different story – more sinister than had been thought – a tale of bigotry, legal chicanery news media manipulation, and the hangings of innocent men. McWilliams claims he was coerced by the Pinkertons to postpone publication of his work until the twenty-first century because they wanted to conceal their part in the tragedy and to conceal the true cause of the violent death of the famous prosecutor of the Mollies, Mr. Franklin B. Gowen.

The within text contains the author's original footnotes by which he seeks to authenticate his claims, as well as Editor's Notes to assist the modern reader.

ANTHRACITE LADS
A True Story Of The Fabled Molly Maguires

CONTENTS

Prologue - An Overview

Chapter 1   Immigration from Ireland

Chapter 2   Gowen's Early Days

Chapter 3   Mining Anthracite

Chapter 4   The Murder of George Major

Chapter 5   Dougherty And The A.O.H.

Chapter 6   The Trial Of Dan Dougherty

Chapter 7   The Gowen/Pinkerton Arrangement

Chapter 8   The Pinkertons' Wild Goose Chase

Chapter 9   McKenna Sent Packing

Chapter 10   Assault On Bully Bill

Chapter 11   Kerrigan Murders Officer Yost

Chapter 12   McKenna "Turns" Kerrigan

Chapter 13   The Murder of John P. Jones *et al*

Chapter 14   The Wiggans' Patch Murders

Chapter 15   The "Showtrials" Begin

Chapter 16   Alex Campbell And His Jury

Chapter 17   Trial of Alex Campbell

Chapter 18   Gowen Triumphant

Chapter 19   Confrontation In The Capitol

Chapter 20   Gowen's Mysterious Death

Epilog

Notes To The Text

Appendix

Citations

Index

Prologue – An Overview

It was a dangerous thing to be a hard coal miner, one of the anthracite lads. The first time my da' took me down a mine shaft he explained, "Matthew, we must face the reasons for our fears, if we're to live with 'em."

Even before I witnessed the hangings of all of those men, I had avoided spending a night in a room where light couldn't enter. In the darkness I'd think of being in a mine gangway, dark, dank and narrow, with countless tons of rock, slate and "gob" poised to come down with a tectonic force that had already crushed ancient forests and swamps into hard, black coal. I had helped to scrape the bloody pulp that was once a man from under one of those falls of the roof. Now, in the dark I imagine the inside of the hangman's hood which hid the last horrible despair of Campbell, Kehoe and Hester who were not guilty, and the bewildered panic of those like Kelly and Doyle who had acted at the instigation of those who were hanging them. Everything I did seemed to be the right thing at the time, but I agonize whether I might have saved some of them had I done something different. No twenty-twenty hindsight comes, and the light is needed as reassurance that for me, at least, there will be a tomorrow.

I'll be damned if I do nothing more, and forever let the villains have the story their own way. They dumped a pile of lies during those trials – lies spread nationwide by the *Shenandoah Herald*, the

Pottsville Miners' Journal, and all the big city newspapers. It was the hottest news story in the United States from the Civil War to the end of the century. There's even a Sherlock Holmes novel which accepts as true the hysterical reports of the press.

What galled me the worst was to read the best-selling book by the arch perpetrator of it all, Allan Pinkerton, entitled *The Molly Maguires And The Detectives*. It's an unabashed tout of Pinkerton's Detective Agency. It claims that one of his operatives, named McParlan, was sent into the coal region and worked undercover for two-and-a-half years pretending to be a member of the widespread criminal gang known as the Molly Maguires. It brags that the detective brought the entire gang, including its leaders, to the just punishment of the law.

But except for a thin veneer of truth, Pinkerton's book is pure bull---. It's not even internally consistent. Pinkerton wrote it two years after the last trial. He had plenty of time to correct the contradictions in his story. Yet his attempts to do it are pathetic. If we'd had his damned book at the trials to cross examine his witnesses with it, we might have saved some lives!

In spite of all that publicity, I must doubt whether anyone will have any knowledge of what I'm writing about by the time it's safe for anyone to read this. I'm certain that the passage of only twenty-five years would not now insulate me and my family and our friends from prison terms if what I am writing were to be made public. God knows, Kehoe was hanged for what had happened seventeen years earlier. So it doesn't pain me very much that nobody will read this for another hundred years. That should be safe. The problem is that, by then, I'll be no more than a name on a tombstone. Whoever reads this will want to be assured that I'm telling the truth. The only thing I can think of is to make citations to the trial testimony and the public records that prove it. I'll also cite the Pinkerton agents' field reports and especially their expense accounts which Mr. Gowen, President of the Reading, inadvertently placed in my hands.

The first thing to explain is – although the twenty men were hanged in large part because they allegedly belonged to an organized criminal gang or conspiracy called the Molly Maguires – that specific organization did not, in fact, exist.

In Ireland when the peasants banded together to resist local oppression they dressed in various disguises, including women's clothes, and they pretended to be followers of fictional leaders for whom they devised fanciful names such as Captain Justice, Captain Right, Moll Doyle and Molly Maguire. This was because non-fictional leaders would have been hung. And so the Molly Maguire name was available to the newspapers in the United States when they attempted to explain the new phenomena of waves of immigrants from Ireland who were perpetrating some improprietous, anti-social, and downright illegal acts. When the Irish in Cass Township rioted in response to President Lincoln's attempts to conscript them into the U.S. Army, even to the point of stopping a train filled with recruits, Benjamin Bannan of the *Miners' Journal* and others called them Molly Maguires. After that, when any robbery, arson, assault, or vandalism could not otherwise be explained it was attributed to The Molly Maguires. This included violence resulting from labor disputes.

It's true, of course, that there were gangs of Irish criminals. Each of the mines had its own local union, and most of these became dominated by the Irish. Union activity was regarded, at that time, as verging upon the criminal. In addition, there was an Irish beneficial and fraternal organization called the Ancient Order of Hibernians which included in its membership not only the ordinary working men, but also many of the labor leaders, and the criminals. To make it appear that the criminal activity was well organized, the Pinkerton detective infiltrated the Hibernians and pretended that its passwords, rituals, and organizational structure belonged to the criminal element and consequently to the fictional boogeyman of the press, The Molly Maguires. The testimony of detective James McParlan and of Reading Railroad President Gowen as to this organizational issue is a masterpiece of studied ambiguity. They tried not to lie, and from time to time each of them admitted the truth. But no defense counsel

was going to convince a stolid Dutch farmer – who had been reading
horror stories about the infamous Molly Maguires in his newspaper
for fifteen years – that such organization did not exist.

Then again, if our Dutchman insists that the Mollies actually
committed all the violence of which they were accused, then he can't
really believe Mr. Pinkerton's boast that by inserting his secret
operative into their midst he put an end to them. Read Mr. Franklin
Benjamin Gowen's presentation to the Pennsylvania legislature in
July 1875 entitled, "A LIST OF OUTRAGES IN THE SCHUYLKILL
AND SHAMOKIN REGIONS." Most by far of his alleged outrages
(which include legitimate union activities) he places in the
Shamokin region. Rightly so. The newspapers reported the arson of
coal breakers all over the Shamokin area during 1875. These coal
breakers weren't simple buildings. They were sprawling clusters of
buildings, stacked six to ten stories high, and given impressive
names. In six months someone, allegedly the Mollies, burned down
the Ben Franklin, the Helfenstein, the Enterprise, the Margie
Franklin, the Burnside, the Centralia, the Wolf; and the Locust Gap,
twice. Those were nine enormous and costly conflagrations.
Shamokin was the region where the Mollies, if they existed, had set
everything ablaze. Yet not a single one of the subsequent Molly trials
had anything to do with these burnings or with any of the
OUTRAGES cited by Gowen. And those kinds of outrages continued
long after the trials. So whoever committed them remains at large. It's
a wonderment to me that no historian has yet recognized how these
facts refute the Pinkerton/Gowen claims.

There were two reasons why the Shamokin Outrages were never
redressed in Court. The less important reason is that the union had a
rule prohibiting violence to persons. Only property was harmed. The
primary reason is that Mr. Pinkerton is woefully mistaken in his
belief that James McParlan successfully concealed his true identity as
a detective until he appeared in open Court to testify against his
cohorts on May 6, 1876. Me and my da' and the Shamokin leaders
had spotted Jim as a detective a full year earlier, in May of '75.

Looking back, it seems we would have done better had we exposed that beguiling viper for what he was. Though Tom Hurley and Mike Doyle surely would have put a bullet in him then, instead of being turned into his accomplices. This is contrary to everything that has been written about the Mollies but it's true, as you'll see from the detectives' reports. I, myself, warned McParlan that he was discovered and delivered to him the notice to get out of the Shamokin Region and to stay out.

The pressures leading to violence in the anthracite fields are not too complex to understand. The coal operators and the miners preferred to keep the price of coal high. High prices meant high profits and permitted high wages. Both workers and operators tried to keep prices high by stopping production of coal whenever it became so abundant that prices fell. Many of the so-called early "strikes" were really collusive work stoppages agreed to between the operators and the unions.

Conversely, Mr. Gowen's railroad made money from carrying the coal regardless of its price. He wanted a steady stream of production. He easily brought the operators under his control with his control over their freight costs. Also, he bought up the coal lands and leased them to operators who were his cronies. Only the workers remained to be dominated. He sought to end the unions' interference with production by creating a link in the public's mind between Molly Maguire violence and labor violence, thereby gaining public support for limiting the unions' ability to strike. He managed to delude himself that under his benevolent control the price of coal could be set precisely right, so that demand for it would increase, and the operators would make a fair profit, and the miners and laborers would make a decent living, and his beloved railroad would forever prosper, and everyone would be happy.

In reality, there were too many collieries. They were capable of filling even the increased demand from low prices while working only six to nine months a year. Worst of all, the price being charged to the consumers did not include but neglected to cover some dreadful human costs – child labor, lung disease, and casualty rates which were horrific. In 1858 the *Miners' Journal* estimated that a colliery employee would most likely be killed or crippled for life in six years.[1] When my Uncle Ed's boys joined the U.S. Army and claimed they were "goin' to fight against slavery" the miner for whom they were laboring scoffed and made a joke that the real reason they were going was because their skins would be safer down in Antietam and Gettysburg.

My da' said that only a man with Mr. Gowen's blinding ego could fail to see that sooner or later the workers had to strike. It was their only way of insisting that those terrible human costs be borne more equitably by all of those who depended upon King Coal.

Da' wrote to Mr. Gowen and in as diplomatic a way as possible, he told him so, the two of them being friends from their early days in Shamokin. Mr. Gowen ceased responding to da's letters after that.

---

1 *MJ* – Dec. 18, 1858. Editor's Note: A comprehensive study of the risks of anthracite mining at that time may be found in *St. Clair*, by Anthony E.C. Wallace.

CHAPTER 1

IMMIGRATION FROM IRELAND

We emigrated in the winter of '47-'48 before the absolute worst of the famine. The Clauchan or farming cooperative operated by our cousins and us outside Ballina in County Mayo was becoming dangerously over-crowded. Less than an acre of arable land to each family. My Uncle Edward had been finding off-season work in the coal mines of Northern England. There, he was told that he'd be welcomed in America. So he decided to take his wife and his two young sons and get out. Our cousins seized upon this as an opportunity to argue that it would be in the best interests of everyone if he took his two unmarried brothers with him. They were Thomas, age 28, and my da' Patrick, a year younger. In spite of the initial shock of being asked to leave, the brothers' departure was amicable enough. The Clauchan purchased their passage, paid them some traveling money, and promised to care for their old folks for life.

So, that March, when they arrived at the port of Philadelphia, they were admittedly "fresh off the boat" but they were no Micks. Unlike the rush of starving refugees who would soon come after them, they had good health, an education at hedge schools, money, and decent clothes. Not only did they speak English, but each could drop his natural brogue and force himself to talk like an English dandy.

In those early days, it was common wisdom in America, imported from England, that if you were met on the road with a slouching, lazy, dirty, bedraggled, witless and loutish sort of fellow, he could be identified, with no doubt, as Irish. Any attempt on his own part to disguise his nationality would be as hopeless as trying to disguise a pig in a clean shirt. Throughout the 40's and 50's every newspaper, even the most respectable, carried its column of jokes and droll stories featuring the nation's most colorful denizens – the Negroes and the Irish. Though, come to think of it, abolitionist sentiment was spreading and generating sympathy for the Africans, which briefly threatened to disadvantage them as objects of ridicule.

Ed and Tom were in the hold of the ship hauling out the family's trunks. My da' was on the quay with Aunt Ceil and her sons, and they could not avoid overhearing as a group of dockworkers were being entertained by one who was reading the joke column of a local newspaper in a loud voice. The fellow was seated on a keg of nails selecting only the jokes about the Irish.

"Here's one!" he exclaimed and then read, "A muscular specimen of a man from the Emerald Isle swaggered into the counting-room of one of our Water Street merchants and slammed down his shillelagh on the manager's desk. 'The top of the mornin' to ye, sir! I've been told ye're in want o' help.' The manager shook his head negatively and told him with mercantile gravity, 'I've but little to do!' 'Then I'm the boy for ye's!' proclaimed the Irishman. 'It's but little I care about doin' – shure it's the money I'm afther.' "

The man's shoulders shook along with the paper he was holding, as he laughed at the punchline. His four companions chortled their appreciation. Then he continued reading as Ed and Tom put down the second of the trunks containing the family's belongings. They stood to listen.

" 'I got arrested again last night' says Pat, and Mike asks him, 'How can that be? Afther I carried ye from the tavern and dropped ye off at the foot of the stairs, I told ye to take off all of yer clothes and sneak up quietly. How could ye get arrested doing that?'

'Because, ye idjiot, ye dropped me off in front o' the police station!' "

As the dockmen were racked with raucous mirth, Uncle Ed

whispered angrily to my aunt, "Take the children out of sight, will you Ceil? For the three of us are going to teach these yapping curs some manners!"

"No, love. No. Let's just leave!" she insisted. Though angry herself, she was fearful.

Uncle Tom grabbed them each by the arm and argued, "Look at the children. They understand what is being said. We can't let this pass without comment!"

The loud man launched into a third reading which the family paused to hear in embarrassed silence.

"Here's one about Irish pugnacity," he said. "Pat comes into the barroom after a long absence and he is trying to get Mike to bring him current with all the local news. 'Your father and his neighbor O'Toole – they were always fighting over one thing or another. Are they still at it?'

'No, after twenty years of such nonsense, they've finally stopped,' says Mike.

'So they buried the hatchet, did they?'

'No, begorra. They buried O'Toole.' "

While listening to this reading my da', who like the others had disembarked in his newest clothes, reached into one of the trunks and pulled out his best hat. He also took out a fancy gentleman's cane containing a hidden sword, which had been in our family for generations.

"Stay as you are. I'll take care of this." He whispered to them. "Play along."

The entertainer began to read still another story, "Pat and Mike stagger into their church and enter the confessional . . ."

But my da' interrupted.

He sauntered over to the group in his new suit, his hat set at a jaunty angle, and with one hand in his vest pocket and the other holding the cane, he put the stick on top of the newspaper and lowered it insolently to obtain the man's attention.

"I'm sorry to interrupt, my good fellow!" He apologized in the clipped High-English of the nobility. "But would you and a chum be interested in earning an easy shilling each?"

The fellow was startled. At first he seemed irritated by the intrusion, but impressed by da's fine clothes and his casually confident smile, he shrugged and answered, "I suppose we would. That's what we're here for."

"Splendid! Splendid!" da' continued his masquerade. "First, I must inquire as to the location of the offices of the Lehigh Coal Company. It is well known, I understand."

"It's in town, about a mile or so from here," answered the man.

"Splendid! Splendid!" da' exclaimed. "Select one of your chums and take up our trunks and lead the way. Let's be lively!"

The jocular man and his companion did as instructed. Carrying the trunks, which were very heavy, they led the family away from the docks and through the streets of the City of Brotherly Love that were crowded with vendors and shoppers and people on every kind of business. Everyone engaged in the hustle of purposeful activity and nobody paid any notice to the family, even though they were so wide-eyed with wonder that at any moment they might have scraped a cobblestone to test if it was really made of gold.

At the Coal Company offices my da' directed his group down an alley and ordered the laborers to rest their burdens on a loading platform. When they had done it, the two men approached him for their pay but he forestalled them with an admonishing finger and announced, "First, we must all thank God for our deliverance."

Forming a circle around the two men, the family bowed their heads and blessed themselves, and as my da' intoned the prayer, he allowed himself to lapse into his normal brogue.

"Jesus, Mary and Joseph, thank ye fer bringin' us safe and sound to America. Keep in yer care those who we've left behind in Ireland.

"Protect us here in this new land, and teach us to forgive those who – through malice, or congenital thickness or simply bad manners – have trespassed against us. Amen!"

As the family joined in the "amen" and again blessed themselves, the fright on the faces of the dockmen signalled their realization that they had not only given offense to their employers, but they had allowed themselves to become outnumbered and surrounded. They were stout young fellows, but the three brothers towered over them.

"We ain't lookin' for no trouble!" the reader pleaded.

"Nor have ye found any, lads," my da' assured them and handed them each his shilling. "Our money, at least, is as good as anyone's."

The reader pocketed his coin and said defensively, "They were only some jokes."

"True enough. True enough," my da' agreed with a shrug.

As the men began to walk away, my Uncle Tom warned them, "But there are some of us who will be lacking such a refined sense o' humor!"

*   *   *

That was the family's first encounter with anti-Irish sentiment in America. It was an impediment to their obtaining good wages; quite often proclaimed by the posting of the advice that, "NO IRISH NEED APPLY."

Of course, everyone who wanted to work needed to apply. The family chose to regard such notices as proof that those posting them were willfully ignorant and deserved to be duped.

They applied and they did it under their real name. McWilliams could be English or Welsh or even Scottish. They affected to speak like Englishmen and most took them to be that. After a mine boss had satisfied himself that they were good workers, he'd usually let the issue slide.

But they faced another impediment to employment. One not so easily overcome. Ed had worked in the English coalfields and could pass muster as a miner, but Tom and Pat could not. They had got hold of a copy of some pamphlets about mining in England, which they all read on the boat over, and Ed tried to explain everything he had learned. But if Tom and Pat had dared to undertake a miner's work they would have blown themselves to Kingdom Come or they would have fallen afoul of one of the many other more-or-less spectacular ways that working a mine can kill you. So at first, only Ed would sign on as a contract miner. He'd hire Tom and Pat as his laborers. Miners made much better pay than their laborers. Labor troubles were often between the miners, who were independent businessmen, and their workers.

Once in the mine the men could not be closely supervised. Ed showed his brothers how to drill the hole in the coal and insert the charge of black powder rolled in brown paper, and how to make and insert the fuse or squib of powder which, with luck, would burn long enough to allow the three of them to scurry a reasonable distance away from the blast. The blast would break from the face of the breast the tons of mineral deposits that hopefully would consist mostly of coal. The rock, slate, bony coal and coal dust was cast aside as useless "gob."

Normally, when the miner freed enough mineral in the breast to keep his laborers busy, he would go home and leave to them the sorting of coal from the gob, and the loading of it into the cars. But not Uncle Ed. His was a family venture. He wanted to help his unmarried brothers, and get them out of his house. He helped them to load the coal, then the three of them went about the mine learning everything they could about its workings, and generally making themselves useful. They wanted not only to qualify Tom and Pat as miners, but also to earn letters of recommendation from the bosses that they could use to find employment in mines where there were better pay or positions, or where it was safer.

Their willingness to relocate in search of self-improvement brought them to work in mines located all the way from Mauch Chunk[1] at the eastern limit of the Southern anthracite field, to Trevorton at the western end of the Middle anthracite field. These were the first American boom towns. They flourished in the 30's and 40's due to expanded use of anthracite – it burned hotter and longer

---

1 Editor's Note:Renamed Jim Thorp

and cleaner – and because the railroads came. They were thriving communities like Summit Hill, Lansford, Tamaqua, Mahanoy City, Shenandoah, Ashland, Centralia, Mt. Carmel, Locust Gap and Shamokin.[2]

Make no mistake about it, the brothers were not merely seeking to survive. Their goal or quest or whatever you'd call it was to become rich, like other Americans. In those early days the coal was close to the surface and easy to mine at a profit. Skilled miners were in demand and the pay was good.

By the time they got off the train at Trevorton, in September 1850, they had saved money in excess of what was earmarked to be sent back to Ireland. Tom and Pat were certified miners. Any of the three could operate, maintain and make replacement parts for any piece of mine equipment, and Pat was an accomplished surveyor. Each of them vowed that, as soon as they could afford it, they would purchase land and get out of housing rented from the mining companies. In July 1851, Edward purchased lot 8 in block 131 of the new town being developed by William Helfenstein of New York City, the man who had put together the syndicate that owned the mine. In February, 1852, Tom purchased lot 10 in block 121. Three years later, in January 1855, my da' purchased lot 11 next to it. On March 17, 1855, he and my sainted mother, Mary Flaherty, were married. They planned to build their house on the lot.

Da' was then thirty-five years old. Ma was thirty. Both of them had put off marrying because of hard times in Ireland and the business of emigrating. They had no time to waste. So new life burst forth on the wedding night, and my sister Mary was born nine months later, almost to the day.

In that pregnant spring of '55 there occurred one other notable event – my da' pulled the thorn from out of the paw of the lion, the great man himself, Mr. Franklin Benjamin Gowen.

---

2 Editor's Note: See Map, Center Section, Area of Alleged Molly Maguire Activity.

CHAPTER 2

GOWEN'S EARLY DAYS

Of course, Gowen was not such a great man at the time. He was only nineteen-years-old. But he had so impressed Thomas Baumgardner, a merchant from Lancaster, that when he decided to sell his dry goods business and try his luck in the coal region boom, he put the young man in charge of the Iron Furnace which he purchased.[1] The furnace used anthracite to smelt ore into pig iron. It was located a hundred yards or so across Shamokin Creek from the original St. Edward's Church.

Gowen was the kind of fellow who by natural instinct is compelled to make a big splash in any puddle he walks into. Within weeks of his arrival, everyone was talking about him.

He joined a group calling itself the Shamokin Senate that met in the Odd Fellows Hall and staged political debates in accordance with the rules of the United States Senate.[2] The year previous, anti-slavery Americans had formed the new Republican Party and the Democratic Party was split between its pro-slavery and anti-slavery factions. In Shamokin, as elsewhere in the North, most people were either Republicans or anti-slavery Democrats. Young Gowen thought it would be a test of his mettle in debate, as well as hooting good amusement, if he argued that the South had the right to keep its slaves and expand its territory, and the right to nullify Acts of

---

1 Editor's Note: See Schlegel, *Ruler Of The Reading*, pp.5,6.
2 Editor's Note: See *Shamokin Centennial Book*, pg 20.

Congress which it deemed unconstitutional; all to the better preservation of the National Union.

Gowen's performances at these sessions were entertaining, impassioned and convincing. He kept all of his facts straight. And as my da' later discovered, he had not only gone through elementary school (and a Catholic one at that) but he had also attended the renowned Beck's Academy where many of the scions of the slaveholding Aristocracy were sent for higher education. He had become privy to his Southern classmates' attitudes and premises, no doubt by arguing against them. And because he was so thoroughly prepared, no one could match him in debate.

This was especially frustrating to some of the more emotional abolitionists. Some of them began to feel that they were themselves personally implicated in the sin of slavery because they were failing to defeat the arguments in favor of the evil that so glibly fell from the lips of the upstart from Philadelphia.

One Sunday after mass at St. Edward's, my mom and da' recrossed the Shamokin Creek on their way back to Trevorton when they saw Gowen inspecting the security of the horses and mules which were stabled on the furnace grounds. Da' boldly introduced his wife and himself, suggesting that at some convenient time they would be interested in learning how an iron furnace operated.

The young firebrand was delighted by their interest. His handsome features – high forehead, aquiline nose and strong mouth and chin – glowed with enthusiasm. He insisted on taking them on a tour of the place at once, saying it would be safer and more pleasant to learn about the process while the production line was shut down for the weekend. He was charming and modest. He showed them the furnace, and the engines imported from England that powered the conveyor equipment. He described how the ingots were poured. There was no hint of the aggressive advocate in the good-humored twinkle of his wide-set blue eyes.

When the tour was ended, he and da' both would have been satisfied to part pleasantly with no talk of politics, but mom was not one to avoid confrontation.

"I think it's only proper for me to warn you, Mr. Gowen," she said. "Some of our more ardent Republicans are saying that because of your defense of slavery, you deserve a sound thrashin'."

Gowen had heard this before and was not surprised. He replied, "But I don't really approve of slavery, Mrs. McWilliams. I'm merely playing the part of those who do. I take no account of such threats."

Da' said, "I think people wonder why you choose to do it."

"At first, simply for the fun of it," Gowen admitted. "But now I think it's useful. The Fugitive Slave Act is chasing the Negroes to Canada amid a howl of outrage and the Kansas and Nebraska violence is spreading. Our Union is on the verge of splitting apart. There's no way of saving it unless the South and the North can come to some agreement about slavery. But no agreement can be reached by people who can't even talk about their differences without ripping at each other's throats."

Da' nodded and said, "So making us listen to the Southern point of view is the first step in coming up with a compromise about it."

"Right", said Gowen. "In the North even those who aren't morally repelled by slavery, view it as something which must not be permitted to spread. The South believes that slavery and its extension into the territories is necessary to the preservation of its economy, its culture and its existence."

"So, is compromise possible, Mr. Gowen?" mom asked. "Or is the South correct? Is slavery necessary to its existence?"

"Does the truth matter, so long as they believe otherwise?" Gowen answered.

Da' observed, "The South fears something else as much as it does the abolition of slavery. It fears democracy and the tyranny of the majority."

"Very perceptive!" Gowen exclaimed. "When a majority becomes too self-serving, it can be as oppressive as any dictator, and with similar results. Zealots make bad democrats."

"The French Reign of Terror proved that point," said da'.

Gowen nodded in agreement, "We had our own experience in Ireland with people who resisted democracy because they feared what a majority might do."

"Oh! You're Irish!" da' exclaimed. "I thought you might be, from your name. We're from Ballina, County Mayo."

"My father was from Newtownstewart, County Tyrone," Gowen reciprocated.

"Only seventy miles separated us," da' noted. "But the religious rift was much wider."

"Not in our case," Gowen objected. "We were Protestant, but also United Irishmen. We believed that the only religion not to be tolerated was the one which wouldn't live in peace with the others."

"Amen to that!" da' approved.

"But I also believe, Mr. Gowen," said mom, "that slavery is another thing that can't be tolerated."

"Why not, madam!" Gowen demanded, seeking to test her reasoning. "As a historic fact, we have tolerated it to this point. Why not continue!"

Da' interjected, "Because this country is no longer a federation of diverse States. It's becoming one. With common language, religions, customs, trade and business. Tied together by ever-expanding networks of iron rails. And slavery isn't a private affair. I can't own slaves if they can be free in my neighbor's yard."

"But most important of all," mom insisted, "it offends humanity. And too many of us have become too humane to endure it."

"That puts the case as well as ever I've heard it, Mr. and Mrs. McWilliams," Gowen said. Then he pleaded with an impish grin, "But don't spread the news that I agree with it."

* * *

Later that week on Friday, the sixth of July, the Shamokin Senate held a special session outdoors. Spectators had been over-crowding the usual assembly room. Benches were set up in the Market Street mall. Most of the town attended, including mom and da'.

Gowen, playing the part of South Carolina's John C. Calhoun, was spectacular.

"By overpopulating itself the North has upset the balance of power on which this country was founded!" he cried. "Northerners slyly got us to agree to stop importing slaves, but then they went out and scoured all of Europe to import cheap labor and increase their voting majorities. They have imposed tariffs on imports to protect their own manufactories, which causes everything needed by the South to be more expensive."

He had mastered the vitriolic rhetoric of a plantation owner defending his rights above all others – saying slavery was none of the North's business – a done deal in any event, and a concession by the founding fathers that couldn't be reversed.

And then to finish, Gowen went on the offensive with relentless logic:

"Did not the States exist prior to this Federal Government? Is not that government the creature of the sovereign States that created it? Who dares to say that the Creator has not the right to nullify the willfulness of his own creature? The very Scriptures refute them!"

The debate raged for hours. When it was ended, a knot of spectators attached themselves to Gowen with some few congratulatory remarks but mostly with questions, comments and arguments. The discussion threatened to go interminably. Da' sensed that Gowen had tired of it. He provided the young man an opportunity to break free from the group.

"Sorry to interrupt, Mr. Gowen," he said, "but it's getting dark, and Mary and I have five miles to Trevorton. If we're to walk with you as far as the furnace, we should leave now."

It was common practice for the furnace manager to inspect the plant premises each evening.

Gowen smiled with gratitude for da's assistance and said. "Quite right, Mr. and Mrs. McWilliams. Excuse me, ladies and gentlemen, but I must be going. Good evening to you. Evening."

He permitted mom and da' to lead him off the mall to the west

along Arch Street where, out of earshot of the crowd, he thanked
them for rescuing him.

"I was impressed by your performance, Mr. Gowen," said mom.
"But even knowing it to be a performance, I couldn't take pleasure in
hearing those arguments."

Da' added, "Your opponents who were stung by the cut of your
wit, took no delight in them either."

They walked three blocks to the bridge that spanned Carbon Run
before it emptied into Shamokin Creek. There, waiting for them,
were some abolitionists who had been outraged by Gowen's wit.

Three were on the near end of the bridge, blocking access to it.
The man in the middle slurred his menacing words.

"We're here to put a shtop to your Souvern insolench!"

His cohorts seemed drunk also, but nevertheless dangerous.

Mom and da' and Gowen had the creek to their right. They turned
around, but from behind a pile of lumber being used to construct a
structure on Third Street, five more toughs stepped out to complete
the well-planned ambush.

Their leader announced, "It's time you learned that there is a price
to pay for slavery, Mr. Gowen!"

The young man was frightened but determined to be brave, and
with clenched fists he retorted, "I have never owned any slaves and I
never will!"

Mom called out, "He's only pretending to take the part of the
South, for purposes of debate!"

The leader tipped his hat to mom and said, "You and your
husband may return to the mall, madam. Our issue is with this
copperhead."

The five men moved aside to allow mom and da' to depart.

"But gentlemen!" da' objected in a tone which pleaded for their
understanding. "You're lettin' yer hatred o' slavery to overcome yer
sense o' decency and fair play, and yer ability to distinguish words
from deeds - in short, yer reason. Enough of this! Lay off!"

Some of the men shook their fists at da', and Gowen grabbed him by the shoulder and pleaded, "Go, Mr. McWilliams! I don't want to see you and your wife suffer for what I've done. Please leave. I prefer to deal with this alone, please!"

"Do as he says, Mick!" the leader growled threateningly.

Da's eyebrow arched in reception of the term "Mick". But he twirled his gentleman's walking stick, weighted with lead at both ends and hiding a short sword. He smiled reassuringly as he bowed to young Gowen.

"Then, my wife and I will take our leave from whence we're so universally not wanted. And we bid ye, sir, 'bon chance' until 'au revoir'."

He winked at Gowen to reassure him that he would not desert him, and the attackers suspected as much as well.

They watched suspiciously as da' slowly and deliberately escorted mom through their midst, but then stopped a mere ten paces away. With his back to them, he ceremoniously handed mom his high hat and his suit coat, and she took two quick steps away from him and stooped to pick up something off the ground.

Da' then whirled gracefully to face the abolitionists. None of them were familiar with Irish faction fights, nor with the havoc that could be wrought by a weighted club in the hands of an accomplished stick fencer.

The leader rushed at da' and received a thrust of the stick in the groin. It doubled him forward, and da's knee shot up and propelled him onto his back in the dirt.

Two others charged from either side of da'. One received a forehand and the other a backhand swipe of the stick on their temples, sending them reeling and holding their heads.

Gowen grabbed one of the assailants around the neck and grappled with him on the ground.

And mom had not been standing idle. She had picked up a fist-sized rock, removed one of her stockings and put the rock inside.

When the fifth man grabbed da's stick and was tumbled over da's back to land at mom's feet, she thwacked him with her slinged weight, using precisely the force required to render him unconscious.

Seeing their larger number so easily defeated, the three drunken blockers of the bridge stumbled away, and Gowen let go of his man who also ran.

Mom handed da' his coat and hat and started across the Arch Street bridge saying, "Let's get out of here before those thugs revive."

Walking the four blocks to Gowen's furnace, she and da' tried to convince him that he should stop antagonizing the abolitionists.

"But they shouldn't feel the way they do," he objected. "We clearly announced it's only a debate. We're merely presenting the arguments of the sections we're supposed to represent."

"You're too convincing, Mr. Gowen," da' told him. "Some people will never understand how you can be so convincing without actually believing in what you're saying. You're by far the best at it that ever I've heard."

Even back then, Gowen was stubborn. He refused to end his participation in the Shamokin Senate.

"But I'm not so foolish as to think this can't happen again," he told da'. "If you'll be my escort to here and back to my hotel after each debate, I'll pay you for your trouble."

Da' thought about it and said, "I'd do it, but it's the five mile walk to and from Trevorton that's the problem."

"That's no problem," countered Gowen. "We'll loan you one of the foundry horses."

Da' jumped aboard the deal which gave him the use of a horse on Fridays, Saturdays, and to go to church on Sundays. And that's how the great man became a friend of the family.

* * *

He was in Shamokin almost three years. I was born near the end of the last year, and he might have been my Godfather were he not Protestant. They were formative years for him; a happy time in which he courted and won the heart of the young woman from Sunbury, Esther Brislen, who became his wife. The engaged lovers regarded mom and da' as their elder confidants. After they were married and had moved to Pottsville, it was Esther who began writing to mom. Then Gowen wrote to da' saying he wanted to keep current with the Shamokin area from a miner's point of view.

But most of what he wrote in the letters to da' was personal. He lost his shirt trying to operate a mine. On the brink of financial disaster, he advertised for new partners. Against da's advice he stipulated, right in the ad, his rule that the men were to be paid in cash only; not in script on the company store. No investors responded. The property was sold by the sheriff and he was left with large debts unsatisfied.[3] Undaunted, Gowen studied law and he hung out his shingle in Pottsville in 1860. That was the year before the Civil War.

Gowen's father was strongly against the war, claiming that slavery didn't cause it, but rather it was a plot by Lincoln and the Republicans to embarrass the Democrats. Nevertheless, Gowen's younger brother, George, who he had brought to Pottsville to work with him, was one of the first to enlist in the Union Army. Gowen himself made a beautiful speech, published in the *Miners' Journal*, which adopted mom and da's position that a compromise with slavery was no longer possible.[4]

Had Gowen enlisted in the army, he could not have made the payments that he had promised to his creditors. He stayed in Pottsville and practiced law. He was elected District Attorney of Schuylkill County on the Democratic ticket in 1862, and with that salary, on top of a thriving private practice, he worked himself out of debt. His name was drawn in the draft of 1863, but by that time he was the busiest and best-paid lawyer in the County, and for that reason he couldn't afford to go. Also, he had small children. He paid the fine and stayed out. So fast did his private practice grow that to keep up with it, in 1864 he resigned as District Attorney.

---

3 Editor's Note: *St. Clair*, pg 404
4 Editor's Note: Schlegel, pg 9

The winter of 1865 came with horror. His sons, James, age six, and Franklin, age two, contracted an illness that baffled the doctors and they both died. It was the worst time of his life, he wrote to da'. His Esther was inconsolable. Then, on April 2, seven days before the war ended at Appomattox, his brother, George, whom he dearly loved, was killed leading a charge against the Confederates.[5]

Da' and mom took the train to Pottsville to offer their condolences to Frank and Esther on these occasions. Da' said that Gowen had changed. Something in him was missing. The wit, charm and intelligence were as before. But where there had once been a cheerful and careless optimism that things would work themselves out, there was now an invincible determination that things must bend to his will. He would place no trust in what he himself did not control.

He became head of the legal department of the Reading Railroad in 1866 and its acting President in '69 and he relentlessly drove it to dominate the coal business which was the prime source of energy for the nation's industrial expansion. He became more of a celebrity than Carnegie, Rockefeller, Morgan or Vanderbilt. So it's possible you've heard of him. You would have, back then.

---

5 Editor's Note: Schlegel, pg 10

Chapter 3

Mining Anthracite

I didn't meet Gowen myself, except as an infant, until April of '75 when I was eighteen and out of the mines for over a year. To explain how it was that my da' talked me into going to Philadelphia to visit him and ask for his help, I have to confess that I failed as a miner; couldn't force myself to work underground, though God knows I tried.

My brothers and cousins managed to do it, leaving me the scandal of the family, though my da' insisted there was no shame in it.

All the boys are taught "readin', writin', and 'rithmetic" until the age of eight or nine then they're sent to the colliery to continue their education and begin their life's work as breaker boys. The breaker[1] is a very tall assembly of buildings, to the top of which the coal is hauled direct from the mine, so that gravity can assist in chuting it through a series of iron bars, crushers, roller screens and cleaning jigs and onto the long, picking chutes where the youngest of boys and the oldest of men remove any remaining slate or other gob by hand. These young and old arbiters of last resort are the breaker "boys". It's a tedious job at best, and at worst we lived in fear of the stick carried by the overseer. The uncleaned coal crashed loudly as it was dumped from the cars into the top hoppers.

There was clanking and grinding from the crushers, clatter from the churning roller screens, and whirring and pounding from the

---

1 Editor's Note: See picture, Center Section.

engines, and screeching from the conveying machinery powered by them. All that and the roar of the coal sliding down the sheet iron chutes was a continuous din around us. Any coal that had been saturated in sulfuric mine water stunk like a drunkard's fart. Worse was the black dust from the crushing and falling coal; a choking grey cloud that engulfed us, coating over the forty glass windows in the walls and making a mockery of their efforts to allow daylight to enter. It was ten straight hours of being digested in the bowels of a flatulent mechanical beast.

But the structure which enclosed this dirty, stinking, cacophonous operation was immense. There was no feeling of being enclosed. I could do the job, and I worked at it from when I was eight until I was twelve. Then I became a "nipper" and my trouble started.

The nipper's job was important, but it was easy work, and during our years of sorting out slate in the breaker, we looked forward to graduating to it. You simply sat beside a tunnel door and opened it to permit mine cars to pass through. If you didn't open the door on time the cars would smash into it, derail and spill coal all over the place, and shut down that section of the mine. But it could be worse if a door somehow got stuck open. A proper mine is ventilated by sucking the bad air out and drawing fresh air in. This means the inflow and the outflow must be kept separated. So parallel to both sides of each working tunnel are dug smaller ventilation shafts, one for air in and the other for air out. The tunnel itself is closed off with doors to prevent too much of the separated airflows from being pulled into it. If that should happen, pockets of mine gas or airborne coal dust accumulate in the working breasts or other, outlying areas, and things are likely to be blown to hell.

When the inside boss at the Trevorton took me underground for the first time to post me at my tunnel door as a nipper, I was full of self-importance. Then he left me alone. There were no lights, save the small oil lamp set on the peak of my cap. As I sat on the crude bench my mind became filled with hitherto unimagined terrors. The darkness and solitude clutched at me. The cap lamp glowed out only a dull five feet, so it was like being buried in a ten-foot circle of blackness. Only the lamp's faint reflection off of the iron car rails at my feet was what saved me from total panic. The gurgle and dripping of water surrounded me. I put out my hand to feel the cold and wet, unyielding cut of the tunnel. I reached up and felt the roof. I imagined it coming down. There was the smell of growing fungi, carbide gas, sulphuric acid and rotting mine timbers. I thought I couldn't breathe. From a distance came a low, rumbling noise. I had heard of crews cutting into worked-out, water-filled mines. I knew it wasn't likely in Trevorton, but you could drown in a mine. I thought I felt a vibration, and listened more intently. There was a shuffling and a scratching. It was rats! Most miners felt safer if rats were around, but I loathed them.

"Oh, Christ!" I prayed for some of the cars to come to me.

Most of the door boys who were killed were those who fell asleep and were crushed by broken doors, or trampled on by mules, or they were struck by runaway mine cars. I was in no such danger because the pounding of my panic-stricken heart kept me awake. But I was developing a morbid apprehension of being in a closed place, and this threatened to put the sprag to my becoming a miner.

I fought the feeling. I talked the underground boss into taking me off the door and letting me be a "spragger". That was dangerous work. But mostly it was done where the gangway was less confining – somewhat lit by lamps and doublewide in places. The widening was to allow empty cars being pulled up a grade to be passed by loaded cars that were sent downgrade under their own momentum. The loaded cars were piled high and heavy, and if gravity went unchecked, they would jump the track or spill a lot of their top coal. We slowed them by sticking logs or sprags – two feet long and three inches in diameter – into the wheels. The first cars in any train were easy to sprag, but as the trip gained speed the spragging of the last half of the cars required increasing dexterity. Sometimes we chased the last cars, sprags in hand, trying to insert them. You had to be nimble and quick and able to squeeze between mine cars and between the cars and the "rig", which is the wall.

At thirteen, I had a spurt of growth that knocked me out of that job – literally. Chasing the cars, and unaccustomed to ducking, I ran my head into one of the collars or top braces. I lay with a concussion on the upgrade tracks. Both of my legs would have been cut off, except that the mule pulling the empty cars upgrade was an intelligent one. She saw me there and refused to go over me in spite of the curses and cracking whip of her driver.

Well, then and there I fell in love with that mule. I made it my business to become her driver, and there wasn't a bit of sarcasm in my heart when I sang to her the driver's ballad:

> My sweetheart's the mule in the mines,
> I drive her without reins or lines,
> On the bumper I sit,
> Chew tobaccy and spit,
> All over my sweetheart's behind.[2]

At fifteen I'd grown to a gangly six-foot-four, and I was not content to be doing a boy's work at a boy's pay. I had no problem with the physical part, the carrying of the picks and shovels, and axes and drills, and bars and lumber, and sledgehammers, and powder and fuses for a mile or more along the gangway; and the setting of the heavy braces to support the roof at the workface, and drilling and firing the blast hole; nor with breaking the coal from the gob and shoveling the finer stuff and hand stacking the larger pieces into the car. What I couldn't do after a while was to force myself to go up the manway, which was dug into the steeply sloped coal seams to give access to the working face of the breast.

Imagine yourself in the pitch-blackness of a six-by-eight-foot tunnel a thousand feet below ground. You plod into the darkness of the tunnel for more than a mile from the shaft giving access to the outside. At the end of it, you grope overhead to find a thirty-inch square sloping chimney, going up. It menacingly reaches into even

---

2 Editor's Note: See picture of a mule boy, etc., ante. Center Section.

more confined darkness for a couple hundred feet or more. You must go up that chimney on rough wooden rungs carrying all that equipment and lumber, because up there is the working face. I did it for almost a full year before the fear closed in on my mind and I could no longer force myself.

I still burn from the shame of it although, as I said, my da' claimed there was no shame. But all that aside, what was I to do for my life's work?

* * *

Much had occurred in my growing years. My sister Mary and I were joined by two brothers and a sister. Uncle Ed was killed when a flash of firedamp exploded his powder keg and his family moved to Ashland. The old Trevorton had burned down in 1865. Though there were other mines, there was no work to the west of town since the coalfields ended there. Uncle Thomas moved his family to Locust Gap. In 1871 we moved to a big, half-acre lot on Tioga Street on the east side of Shamokin, commonly called Springfield. There we built a detached house complete with a coal furnace and hot water heat and a nice stable. It was a strategic location, less than a mile from the new St. Edward's Church on Rock Street, fairly close to Uncle Tom and our cousins in Ashland, and within working range of two dozen collieries. We were never without a couple of horses and a serviceable buggy. During the war and the postwar boom, we had prospered at coal mining, and I needed an occupation that was as good, or better.

Da' thought I should read law and become a lawyer. I worried that I wasn't smart enough. If I weren't, then all that time and effort would go to waste. Besides, I didn't know if I'd like it. Da' insisted I try it. He wrote to Mr. Gowen asking what I should do to prepare myself. Gowen made up a list of a couple of hundred books for me to read. He sent me a trunk containing over forty of the titles with instruction that when I had mastered all of the books, I should return those in the trunk to his home in Mt. Airy, in person, and he would examine me and render his opinion whether I should continue. Well, how in holifernes could I refuse to be guided by one of the top lawyers in America, who was also the most prominent businessman of his time? Da' was full of suppressed self-satisfaction, and I wonder how much of Mr. Gowen's plan came from him.

I had to be employed while I was reading all those books. I was offered a job as assistant to John Siney, the President and founder of the Workingman's Benevolent Association. Da' was a staunch union man, but he cautioned me that taking the job might cost me the goodwill of Mr. Gowen. The depression of 1874 had started and it was clear that the union was going to collide full steam ahead with the Reading Railroad and its iron-fisted President.[3] Because of the depression the mine operators wanted to reduce the wages paid their workers. But the wily Gowen wouldn't let them. He saw there was no stockpile of surplus anthracite to cover a long work stoppage. A strike would force the users of hard coal to convert to the use of soft (bituminous) coal and they might be lost as customers. So Gowen forced the operators to pay the higher wages and to stockpile the coal through the end of the year.

When the coal companies' wharves and storage yards began bursting with piles of surplus coal, it became obvious to everyone that Gowen and the operators were determined to force a strike and break the union in '75.[4]

Nevertheless, I became an organizer for the W.B.A. I was a true believer in its message. We talked to the men about the prospects of the anthracite trade, listened to their complaints, and suggested what laws were the most needed. We argued that the survival of the union itself was more important than any number of wage or safety issues. Because only through union solidarity could we win any of what we so sorely needed.[5]

When Gowen found out about my job he wrote my da' that he

---

3 Editor's Note: See Schlegel, p. 63
4 Editor's Note: See Schlegel, pg. 64
5 Editor's Note: See *St.Clair*, pp 396-398

approved. He said that he and my union's President had managed to develop friendly personal relations in spite of their differences about business matters. I thought that was very gracious of Mr. Gowen. I don't think so now.

It was on a visit to Mahanoy City as a W.B.A. organizer that I first ran into the Pinkerton/Gowen version of the Molly Maguire conspiracy, though I didn't recognize it as such at the time.

Chapter 4

The Murder Of George Major

Workingman's unity was a difficult concept to sell in Mahanoy City because, like other anthracite towns, that community of five thousand souls was fundamentally divided. The mine owners who leased to the operators lived in New York or Philadelphia or, if local, in baroque mansions aloof from everyone. The divisions among classes of employees – inside worker, outside worker, mechanic, miner, laborer, the boss and the bossed – were often as pronounced. The grocer, the butcher and other merchants often sided with the workers who typically owed them money, and against the operators who were often late in making payroll. All groups depended upon but resented being dominated by the Railroad and its Coal & Iron Company which had their own private police force.

The people spoke different European languages and different dialects of English. They were of different religions, or of different versions of the same religion. They followed separate customs and traditions, and generally regarded other ethnic groups with varying degrees of ridicule, unease, suspicion, dislike and hostility. They referred to others, and often to themselves, as "Limeys," "Orangemen", "Micks", "Krauts", "Hunkies" and "Wops". As often as not, they intended merely to be descriptive and not to give insult. That depended upon the context, and upon the tone of voice of the speaker, and the skin thickness of the listener.

By far, the most dangerous division was caused by the

overwhelming influx of those who had fled the Great Famine and
its aftermath. They had arrived in this country diseased and
malnourished, accustomed to chronic unemployment, uneducated,
despairing and desperate. And yet they were of such numbers that
the established English, Welsh and Germans feared they would take
their jobs, and replace them as town councilmen, burgesses and
constables.

There was a deadline down the middle of Main Street in Mahanoy
City. You couldn't see it. But people referred to it as the deadline and
everyone knew it was there.[1] After dark, the English, Welsh and
Germans wouldn't be caught dead on the Irish side of it. Nor did the
so-called Mollies dare to stray into the other turf. Gangs of English,
Welsh and German tavern toughs roamed the place. Most pegged
themselves with a sinister sounding name out of Indian lore – the
Modocs. The areas were so antagonistic that each had its own
separate fire department.

I was at the Emerald House on Main Street, on the Irish side of the
deadline, that last night in October of '74. My sermon on the gospel
of W. B. A. unity which I had delivered in the upstairs assembly room
had been fairly received.

"You get the same pay for the same work as the English and the
Dutch," I told them. "You die alongside them from the same faulty
ventilation. So in that, at least, your interests are the same."

"That may be true," a man had countered, "but when work gets
scarce and the Welsh boss assigns all the available breasts to his
Welsh buddies, then the spit hits the fan."

"To take a breast away from a man who is properly working it is
like stealing his money in the bank, as well as his wallet," I had to
agree. "But if everyone belongs to the union we can prevent that."

Afterward they invited me downstairs for a pint. We sat at a table
in front of large windows that ran the full length of the barroom, and
we could see outside that it was promising to be a lively night.

---

1 *MJ* – Ap. 24, 1875, test. M. Bowman and *MJ* – Ap. 29, 1875, test. J. Foley.

Halloween is the evening before winter in the Irish calendar. It is Pooka Night, when the earth comes under the sovereignty of the supernatural and the spirits roam. There is feasting and partying and merrymaking and the playing of games and dares and pranks. This was one night when the bravest of the Irish lads defied the deadline and the gangs of Modocs. So the English and German householders suffered their share of the foolishness. A group of the scamps who were older than most ventured into the crowded barroom. They were panting for breath from running, and it was obvious they were in hiding.

Sitting next to me at the table was Mike Clark, proprietor of the Emerald House[2] and he demanded of them, "What's up?!"

"Please sir, let us stay here for only a little while 'til we see whether the Chief Burgess and his brother Jesse are on the prowl for us," pleaded the leader of the boys.

"You mean to say you've run afoul of the authorities?" growled Clark.

"Not exactly!" came the answer with a big, gap-toothed grin. "I'd say, rather, it was Jesse Major who put his foot afoul of us!"

The boy and his buddies burst out laughing as though something witty had been said, but Clark cut through their raillery by slamming his meaty palm on the table.

"What about Jesse Major?!" he demanded.

"We gave him an intelligence test," smirked the lad. "We got a big glob of fresh, wet pigplop – ye know that stinks the worst of any – and put it in a paper bag. After putting a match to it, we placed it on Jesse's front porch and banged on the door. Well, he comes out . . . in his stocking feet, I swear to God . . . and he tries to kick the burning bag off his porch. His foot rips through into the plop, and the bag gets stuck on his foot, and he starts hopping around on his other foot, and he falls on his ass bellowing curses!"

The boy and his friends doubled over with mirth, which was

---

2 Pinkerton, pg 306, 307

infectious, and everyone at our table joined in. Finally, Clark
regained his composure and said, "Then what?"

"We couldn't stop from laughing out loud, and so Jesse sees us
and charges after us, stocking feet and all. Ye know how mean he is.
He'd 'a killed us, no doubt, had he caught us. He screamed out that
he knows who we are, and he's going to get his brother, the Burgess,
and come after us. We ducked in here to hide."

Clark once again slapped the table, this time with approval,
saying, "Good work, lads!"

He gave them some pennies to buy themselves a treat. But before
they left he warned them, "Ye best go to yer homes and stay there.
I'm not worried that our Chief Burgess will get ye. He's decent
enough. But his brothers, Bill and Jesse, are a different matter; they're
something worse than just mean Welshmen. If they get their hands
on ye, this could become serious. Ye hear?!"

In company with me and Mr. Clark at the table was one Dan
Dougherty, who hailed from Centralia. I'd met him there several
times at picnics of the St. Ignatius' Church.

Dan was above average height, five-feet-eight or so, and broad
shouldered, with regular features.[3] Upon moving to Mahanoy City he
had become prominent in the Humane Fire Company. He was so
proud of the fact that he always wore the dress parade cap of his
outfit. He had a sympathetic regard for a young stranger in town,
such as myself.

"My buddies and I have been invited to a party, Matt," he said to
me. "Why don't ye come along with us?"

"I wouldn't want to impose," I said, by way of accepting the
kindness.

"Nonsense, Matt," Dan smiled. "The girls who invited us have
given us standing orders to bring along any strangers who aren't riff-
raff. I presume ye'll qualify!"

I laughed at his jibe and retorted, "In that case, ye can also

---

3 *MJ*-Ap.27, 1875, test. Mrs. Emma Holt. *MJ*-Ap.28, 1875, test. M. J. Salmon

presume my thanks!"

We left the Emerald House and went to a large barn on the Northern outskirts of the town where the party was in progress[4]. Its organizers had enticed the presence of respectable ladies of the town by assuring them it would be a safe and proper entertainment, and therefore alcoholic beverages were not allowed. At a gate permitting entrance to the fenced barnyard, three attendants were arguing with a tall man wearing a broad brimmed slouch hat, who had tried to enter with whiskey.

The man saw us approaching and beckoned us to hurry, saying to my companion, "Haloo, Dan. Dese boyos air fixin' to shun me out, e'n though I told 'em they cud keep me bottle outside. Ye'd best ha' a word wid 'em, eh?"

Dan Dougherty whispered to me, "It's John McCann. He's a bit of a Mick, but he's a member of my own fire brigade."

"I can't help ye now." He responded to McCann's demand, "This party is not only dry, it's by invitation only. But tell ye what! Go take yer bottle and lave it back at yer boarding house, and when ye return from there clean and sober, then I'll vouch fer ye bein' as good a man as any in the Humane! What d'ye say?!"

McCann scowled and squinted at me as though I might help him, but I said, "He's being fair wid ye, Mister. An' he's givin' ye good advice. Take it."

His air of indignation having been deflated, the drunk shrugged and agreed, "Allright. Dat's what I'll do then. See yez boyos in a bit."

When he had walked away some distance, Dan told the gate attendants, "I doubt he'll bother to come back."

"That's good," they told him, and they admitted us cordially.

The barn was festive with jack-o-lanterns and decorations and lively music, and I enjoyed meeting Dan's friends and dancing with the young women. Shortly before eleven o'clock, he sought me out to tell me that he and some others were on their way to pay their

---

4 *MJ*-Ap. 29, 1875, test. F. Cownan

respects at a wake[5], but they'd return to the party. Suddenly the music and the dancing were shattered by the rasping shrieks of a whistle.[6] Everyone stopped what they were doing to listen, and the whistle shrieked again.

"It's a fire!" Dougherty exclaimed. Without another word he bolted for the door and raced back to town. My instinct was to follow along and help if I could.

We got as far as the hotel where Dougherty boarded and he explained, "I must go in and put on my boots. I'm at the nozzle."

I paused to catch my breath. In a minute or so Dan came back out wearing his boots and, as before, his peaked cap embossed with the emblem of the Humane Fire Company.[7]

In that short time, others of the Humane had unrolled a fire hose which stretched across the railroad tracks, past the hotel, and on towards Main Street.[8]

We followed the hose to an alley behind Main Street where there was a glow of fire and clusters of flames coming from a stable. Flying sparks had already spread the burn to another stable and there was imminent danger of a serious conflagration developing.[9]

Dougherty went into the stable yard and picked up the nozzle of the, as yet, inactive hose. Just behind him on the nozzle came none other than the drunk, John McCann, still in his slouch hat but no longer bearing a bottle.[10] McCann was a bit taller than Dougherty but not as muscular.[11] Other men also tended the hose end. Dougherty led them up to the top of the stable's dung heap as a vantage point from which to quench the fire.[12] It was good he had donned his boots. But as yet there was no water pressure in the hose,

5 T. Sheridan, op. cit.
6 F. Cownan, op.cit.
7 Mrs. Emma Holt, op. cit; MJ-Ap. 27, 1875, test. T. Kaercher. MJ-Ap.23, 1875
8 MJ-May 1, 1875,test., Mrs. H. Koch; MJ-Ap. 30, 1875, test., J. McGuire; MJ-Ap.27, 1875, test., R. Lee
9 MJ-Ap 29, 1875, test P. Ryan
10 F. Cownan, op.cit.; MJ-Op.29 ,1875, test., J. Hughes
11 J. Hughes, op.cit.
12 MJ-Ap.29, 1875, test., James Foley

and the men could only watch anxiously as the fire consumed the structures.

To add to the frustration of the Humane men, the Citizens' Fire Company rushed into the alley from the opposite direction.

I heard someone yell, "The Citizens have the use of a plug at the corner of Centre and Main.[13] They'll have pressure as soon as they make their connections. Go tell our own lads to hurry!"

A large number of the townspeople were assembled on a railroad embankment to watch the firefighting contest, and many were in the alleys and streets and yards surrounding the structures. They let forth a cheer when the first rush of water spurted out of the Citizens' hose, and an equally loud cheer when, only moments later, the hose attended by Dan Dougherty shuddered and swelled and shot forth its quenching stream.

There were no street lamps in that part of Mahanoy City, and the night was dark and moonless [14], but even so, I could see the flames turn into pillars of smoke and hear their protesting sizzle as the fire companies performed admirably. Soon it was clear to all that the fires had been contained and were no longer a danger to the rest of the town.[15] The least tenacious of the onlookers began to return to their homes.

With the threat abated, the attention of some of the firemen returned to their inter-company rivalry. I saw McCann, the drunk, point to where a member of the Citizens' had climbed onto the low, flat roof of the second story of one of the stables. Gleefully, he and Dan Dougherty turned their hose on the fellow.[16] They did it in the spirit of Pooka Night, not maliciously.

The man was knocked down by the spray. He scrambled again to his feet, and with Dougherty and McCann trying to hit him again, he dodged around the roof. Finally, he slid himself over the side of the

---

13 *MJ*-Ap.26, 1875, test., F. Jones
14 *MJ*-Ap. 23, 1875, test., M. Bowman
15 *MJ*-Ap. 28, 1875, test., T. Sheridan; *MJ*-Ap. 29, 1875, test., John Foley
16 *MY*-Ap.30, 1875, test., Jere McGuire; T. Sheridan, op.cit.

roof, hung from the gutter and jumped to the ground.

The men working the Citizens' Company hose saw the horseplay to the peril of their fellow. They retaliated by deftly spraying a loop of water directly at the Humane nozzle bearers.[17] It was only a single spurt, but so accurately aimed that it drenched Dougherty and McCann and the others in the stable yard. They took it with good enough grace, and there the matter might have ended.

But standing next to me in the alley was the big, burley Chief Burgess of Mahanoy City, George Major. I'd been introduced to him and his brother Bill. He was himself a member of the Citizens' Company and he was not amused by the water fight.

He ran into the stable yard and there were angry words between him and the Humane men. I couldn't hear them clearly. Trying to get closer, I tripped over a hose in the dark.

An older man helped me to my feet and ordered me, "Get out o' the way, young fellah. Ye don't belong here."

He seemed to be a foreman or other person of authority, but regardless, I followed him.

He asked the men at the nozzle, "What's going on?" and McCann pointed an accusing finger at Major and lied, "De Welshman here, as bold as brass, is afther tellin' us where to direct our flow."

The foreman hadn't seen the spraying incident and, believing the drunk, he sided with his own men saying, "Is that so?! Well, Mr. Chief Burgess, you have no authority here. And whether you like it or not, I'm pleased to say that we can get along without you very well!"[18]

The Burgess puffed with rage. He took the words as an insult, not only to his person but also to his office. His anger was heightened by the fact that he'd been acting most reasonably to censure some very unprofessional and dangerous conduct, and he was clearly in the right.

---

17 *MJ*-Ap.30, 1875, test., H. Baumgarten; T. Kaercher, op.cit.;
18 Jere Mcguire, op.cit.

He had a bad temper, and as it exploded he screamed, "That rips it! I'm tired of your lying Irish bullshit. You fellows have been having things your own way long enough. But you'll see. You're not gonna get away with it tonight. Not this time!"[19]

Major pushed the foreman in the chest with his open palm with enough force to set him on his behind on the soaking wet dunghill. He was only half the size of Major but he bounced up and challenged defiantly, "Oh yeah! Then what exactly are ye fixin' to do about it?"

Major was further angered by the question and somewhat at a loss to answer it. He stammered, "I …uh … I'll do something. I'll do whatever I please. If I have to use force to keep the peace, I will! You'll see!"[20]

He stomped out of the yard and into the alley where his brother Bill was waiting for him.

It might have been at his brother Bill's instigation that George Major then did what he did. At the corner of the alley a mongrel dog, mostly a hound, was howling. It was a stray that had been taken in by the boys of the Humane as a pet. They had taught it to howl approval of their feats of firefighting. Bill Major had been heard to complain that it was making a mockery of serious work.

George Major vented his rage by pulling his pistol from his coat pocket and shooting the dog.[21] The shot collapsed its hind legs, but it scrambled about on its front legs and screamed. Major shot it again.[22] Nevertheless, it twisted and rolled in the street, persistently struggling to get up. Bill Major then took out a blackjack and hit the dog on the head with it seven or eight times until it lay still.[23]

It was a grisly incident. McCann saw it, and he let go the hose and hopped around the stable yard cursing and shaking his fist in a rage

19 Jere McGuire, op.cit.; H. Baumgarten, op.cit.; F. Cownan, op.cit.; MJ- Ap.28, 1875, test., J. Cunningham
20 Jere McGuire, op.cit., H. Baumgarten, op.cit.
21 MJ-Ap. 26, 1875, test., W. Major.
22 MJ-Ap. 29, 1875, test., M. McGrath
23 MJ-Ap.28, 1875, test., M. Donohue.

fueled by whiskey.[24] The people on the railroad embankment saw it and many of them, anticipating there would be more ugliness, returned to their homes. I later heard it said that fights broke out at this time between the Humane and the Citizens' men, but I didn't see any of them.

But most definitely, all horseplay and good will and all mutual satisfaction with success in fighting the fire ceased. Both companies went about the completion of their tasks sullenly and defensively.

The hoses of the Humane had been laid over the railroad tracks. At the request of a brakeman the hoses were uncoupled to allow the scheduled train to pass. Dan Dougherty's coat was soaked from the horseplay and he and I and some others walked back to his hotel so he could change.[25]

We were at a table having a whiskey when John McCann walked in and joined us. He had been to his own boarding house to change and now he wore a long overcoat, though he still wore his wet slouch hat. Held below the table so no one else could see it, he showed us that he had acquired a pistol – a derringer type with four small barrels that rotated in lieu of a cartridge cylinder when the hammer was cocked – commonly called a pepperbox. It fit easily into the palm of his big hand.[26]

Dougherty scowled angrily and ordered him, "Put that damned thing away! Especially in front of the child there!"[27]

The proprietress of the hotel, Mrs. Holt, had entered the bar with her young son. She and her husband were English but were on the best of terms with their trade, which was mostly Irish.[28]

McCann reluctantly retired the pepperbox to his coat pocket. Dougherty finished his whiskey in a gulp – I suspect he hurried to prevent McCann from ordering one – and he led us on the way

24 M.Donohue, op. cit.; M. McGrath, op.cit.
25 T. Kaercher, op.cit
26 *MJ*-Ap. 29, 1875, test., John McDonald; T. Kaercher, op.cit.
27 T. Kaercher, op.cit.
28 *MJ*-Ap. 28, 1875, test., E. Holt on recross; also E. Holt, op.cit. and Thos. Kaercher in *MJ*, May 1, 1875.

back to the stable to roll up the hose.[29]

In the alley the foreman who had been pushed by George Major into the wet dunghill was looking for him to give him a piece of his mind about his shooting of the Humane's mascot.

Someone said, "The Chief Burgess has gone for a drink, but there's his brother, Bill."

The foreman went up to Bill Major who happened to be talking to the Chief of Police, and said to him, "Bill, your brother is a damned mean man."[30]

The Burgess' brother was a bruiser of a fellow, almost six feet of bulging muscle, and he also had a bad temper.

"You say that again," he threatened, "and I'll kick the shit out of you!"

He made a move toward the much smaller man who said quickly, "I'm not here to fight ye, Bill, but only to tell ye about yer brother. You know me well enough to . . . "[31]

Bill shook his fist and interrupted, "Anyone who says my brother is mean is callin' me the same, and I'll lick him good!"[32]

He cast a scornful eye over those of us who were with the foreman, and over those who were coming from all directions to see what was going on, and he shouted out a challenge to all.

"And I'm standin' here saying to one and all, that I can lick any Irish son-of-a-bitch on the ground."[33]

There were curses and howls of outrage. One man shouted that he'd accept the challenge, but he was so drunk he couldn't get his arm all the way out of his coat sleeve when he tried to take it off, and he was pushed aside.

John McCann advanced upon Major yelling something incoherent – he might have lapsed into the Gaelic – but Dan

---

29 T. Kaercher, op.cit.
30 W. Major, op.cit.
31 J. McGuire, op.cit.
32 W. Major, op.cit.
33 *MJ*-Ap.28, 1875, test; M.Foley; T.Sheridan, op.cit.; *MJ*-Ap.28, 1875, test., T.Ryan

Dougherty stepped between them and tried to speak reason.

"See here, Bill!" he said. "We, none of us, came here to fight anything but the fire. You've got no cause for this. There are men as good as you are in the Humane, and right here on the ground, but . . ."[34]

While Dougherty was speaking, George Major, the Chief Burgess, advanced through the knots of people in back of his brother holding out his dog-killing pistol and waving it at everyone, including myself.

The foreman saw him, and thinking the pistol was pointed at him because of his recent argument with Major, he put out his hand defensively and pleaded, "Ah, George! You wouldn't do that!"[35]

At the same moment, Bill Major attacked those of us who were trying to reason with him. In his hand was his blackjack, and he flailed the weighted leather sling wildly, hitting Dougherty, McCann, the foreman and me, and others.[36] I thought he had broken my shoulder blade, but the pain eventually receded.

McCann was pushed backwards, and he brought up his pepperbox pistol and shot George Major in the chest.[37] I don't know if there were two shots or one.

Gunfire was the signal for most people to duck and run away. I ducked but didn't run.

George Major staggered back a pace or so and exclaimed, "My God! Boys, I'm shot!"

A man with a lamp grabbed McCann by the collar and said, "This is the son-of-a-bitch that shot you!"

Another grabbed McCann by his coattails, and also another, but someone tackled them and dragged away the first man. George Major himself wrapped his big hand around McCann's neck and raised his pistol to shoot him but McCann pushed at the barrel to point it away. The knot of struggling men – one of them holding a lamp – whirled

34 *MJ*-Ap. 30, 1875,test., O. Martin; T. Ryan, op.cit.; T. Sheridan, op.cit.
35 J. McGuire, op. cit
36 T. Ryan, op. cit.
37 *MJ*-Ap.29, 1875, test., Pat Ryan; *MJ*-Ap. 28, 1875, test., P. Elliot

around in a macabre dance, flashing light and dark. Then George
Major wrestled his pistol free of the melee and fired it only inches
away from McCann's head. At the pistol shot, everyone holding
McCann hunkered down and he broke free and ran into the night.[38]
George Major groaned and doubled over, and his brother and others
put their arms around him for support. They staggered with him
toward Main Street. He still held his pistol. I followed along. They got
as far as the boardwalk at the corner of the street. Then the Burgess
said he had to lie down, and he handed his pistol to his brother,
Bill.[39] A man with a torch came and provided light and we stood over
the stricken man trying to decide what next to do.

Toward us from out of the darkness of Main Street came three
fellows. The middle one looked like Dan Dougherty.

Bill Major saw him. He took two steps toward him and raised his
brother's pistol menacingly.

I was opposite in the crowd, too far to get to Major in time, but I
screamed, "No!"

Dan's companions moved off to either side of him. He pulled up
short and began to back away.

Major bellowed at him, "Dougherty! You stop!"

He didn't. He continued to step back. Then he turned and bolted
to get around the corner of a fence and out of the line of fire.

Bill's pistol roared. My impression was that Dougherty was hit. He
raised his hands to his head and bounced into the fence. But he didn't
fall down. He kept going around the corner out of sight,[40] and I
wasn't sure he'd been hit.

Major turned back to the crowd clustered around his brother and
someone exclaimed, "Jeez, Bill!"

"You had no cause to do that!" I challenged him.

---

38 Jere McGuire, op.cit.; T. Ryan, op.cit.; Pat Ryan, op.cit.; P. Elliot, op.cit.; M. Donohue, op.cit.
39 W. Major, op.cit.,; M. Donohue, op.cit.
40 T. Ryan, op.cit.; *MJ*-Ap.29, 1875, test., J. Gurdy, Michael McCarl, James Murphy and
William Hollihan; *MJ*-Ap. 30, 1875, test., Wm. Sharp and C. Metz; M. Donohue, op.cit.

But he was defended by someone who said, "It serves the sons-a-bitches right!"

Bill Major stared at me and complained to his group, "This one's a goddamned Mick!"

Then from down the alley we had just exited, there came the crack and flash of another gunshot. The ball whistled over our heads, and everyone crouched to the ground.

The man with the torch and a fellow with a pistol, who I thought might be a policeman, ran into the alley and there were two more shots fired, but nothing came of it as far as I knew. [41]

They decided to carry the Chief Burgess into a drug store to tend to him. From some of the looks being cast my way, I got the feeling I wasn't welcome there. I went to search for Dan Dougherty.

He wasn't around the corner where he'd gone after Bill shot at him, so I sought him out at his lodgings in the Holt's Hotel. He was lying on his back on the floor in front of the bar, surrounded by an ever-growing crowd of people. His face was swollen from hits of Bill Major's blackjack, I presumed. I couldn't get near him.[42]

Mrs. Holt was holding cloths to the back of his head and rinsing them out in a pan of bloody water. She sent some men out to find a doctor.

Dan's friends whom I had met at the dance were standing together. They told me that a bullet had entered through the back of Dougherty's head and was still lodged in his jaw.[43]

I told them that I had seen the unprovoked shooting of their friend Dan by Bill Major.

"There was no justification for it whatsoever!" I said. "If you need a witness in Court, I'll swear to it! It was bloody murder!"

"Thanks, Matt, but you'd best be careful what you say out loud," one of them cautioned me. "It's not only that ye might end up in a

---

41 *MJ*-May 1, 1875, test., N. Garrett and J. Garrett
42 T. Kaercher, op.cit.;E. Holt, op.cit.; P. Clark, op.cit.
43 *MJ*-Ap. 30, 1875, test., Dr. A. P. Carr.

ditch on the wrong side of the deadline. But if they find out in advance what yer goin' to say, they'll get someone to counter it."

Also present at Holt's were most of the others who had been present when the Burgess was shot. John McCann was sitting at one of the tables next to the stove. His overcoat draped over the back of his chair was bloody, and he held a kerchief to the side of his head. He eagerly showed me his wound. It was a cut over an inch long and a quarter-inch wide surrounded by powder burns, obviously the result of a gunshot.[44] He had his little pepperbox pistol in his hand and he showed it to several people as the one that he had used to shoot George Major.[45]

"Why the hell did you do it?" I asked him.

"Sure, didn't he shoot the dog, an' I figgered I was next!" came the drunken answer.

The more times he retold the story, and the more he thought about it, the more he changed it into a tale of self-defense. He went so far as to say that he would have the Burgess arrested for shooting him!

He began to brag about what he had done, showing everyone his little pepperbox pistol.

Then the son of the Mr. Clark who owned the Emerald House arrived. He accused McCann of stealing the pistol some months earlier, when he was lodged at the Emerald House.[46] McCann stopped showing the pistol.

Those who had been sent out to fetch a doctor returned to report that the physician could not be found. Another doctor was sent for, and pending his arrival Dan was moved upstairs to a bedroom.

There was nothing I could do at Holt's to help Dougherty, and when Clark's son invited me to walk with him back to the Emerald

---

44 *MJ*-Ap. 30, 1875, test. P. Mulvey; *MJ*-Ap. 29, 1875, test. Doctor McDonald; *MJ* – Ap.28, 1875, test, J. Cunningham
45 *MJ*-Ap. 28, 1875, test., P. Clark; *MJ* – Ap. 30, 1875, test., Mrs. McDonald; *MJ*-Ap. 29, 1875, test., John McDonald
46 P. Clark, op.cit.

House, I agreed. By the time we got there the word was out that the Chief Burgess still clung to life, but his wound was expected to be fatal.[47] People were saying that Dan Dougherty also had been mortally wounded – his head practically blown off.[48]

I went up to my room and dropped into bed. I was exhausted, but couldn't sleep. My neck and shoulder were sore. Ripping through my mind like a dragline was the horrible chain of events in which useful competition and Halloween exuberance were transformed by racial hatred[49] into an excuse for mindless and brutal assaults on an animal and two men. I decided that I would stay in Mahanoy City until the Wheels of Justice had been set properly into motion. I would see to it that both John McCann and William Major would be convicted of assault and battery, or of murder if the conditions of their victims should warrant it.

Finally, my overwrought mind lapsed into fitful sleep. I had unsettling dreams of illogical violence. I can only dimly recall the dreams now. But I clearly remember what actually happened that night. None of it had anything to do with any Molly Maguires!

I slept almost until noon. At breakfast, I was astonished to learn that Dan Dougherty was to be placed under house arrest! He had been accused by Bill Major of the shooting of his brother, the Chief Burgess![50]

---

47 *MJ*-Ap. 23, 1875, test., Doctor Bissell
48 *MJ*-Ap. 26, 1875, test., W. Jolly
49 Editor's Note: In those times and for years afterward the different nationalities of Europe were often thought of as different races. See: *Seamus McManus, Story of the Irish Race*; A. Hitler, *Mein Kampf*.
50 *MJ*.-Nov. 6, 1874

Chapter 5

Dougherty And The A.O.H.

I tried to fathom why Dan Dougherty, who I knew to be completely innocent, had been charged with the murder.

The Major brothers and their supporters represented a political faction that was devoted to keeping the Irish in their place. After the fire that night, George Major had gone into a bar and broadcast the boast that, "He had strong arms to keep them down!"[1] Some of the folk in Mahanoy City, especially those on the Irish side of the deadline, accepted this as the sole reason for the arrest of Dan Dougherty.

Then again, the shooting of the Chief Burgess had been in a dark and confusing melee. McCann and Dougherty had been standing together wearing similar clothes – except for their hats – and they were about the same size. Some might not have been certain who had fired the soon-to-be-fatal shot. Their prejudice or other motives might have shaded and reshaped their perceptions.

Dan Dougherty was reported to be "verging upon eternity" from Bill's bullet, so Bill Major himself was facing a murder charge. If Dan was identified as the murderer of the Burgess, then that would give Bill an excuse for having pulled the trigger.

I was in the Emerald House pondering these things when a boy ran in and announced that George Major had signed a statement for

---

1 *MJ*-Ap. 30, 1875, test., H. Baumgarden.

the Coroner naming Dan Dougherty as his murderer. I was astounded. In those days a dying declaration was considered to be almost conclusive proof of its contents. I had told Mr. Clark the whole story about that night; that Dougherty wasn't guilty. Clark voiced his puzzlement saying, "I don't understand. George Major had his faults, but he told the truth. He's no liar!"

I went to Mass, then telegraphed headquarters for permission to stay in Mahanoy City, and returned to the Emerald House. All the talk was of the shootings. Rumors were rampant. People said that half of the police force, including the Chief of Police, were willing to swear that it was Dougherty who shot the Burgess. Things looked bleak for Dan.

Word came in from a woman who had overheard a breakfast conversation between the Mahanoy City correspondent for the *Miners' Journal* and the physician who was attending George Major in his last agonies. The doctor whispered to the reporter, "He told me that Dan Dougherty had shot him and that he had shot Dougherty in return. I wrote down his exact words!"

I ran with this information to Holt's Hotel where those who were conducting Dan Dougherty's affairs were headquartered. His condition was reported as unchanged – still critical. I gladly donated a dollar to the fund they'd set up to pay either his legal fees or his funeral bills, as the case might be.

"You have got to convince everyone that it was Bill Major who shot Dan," I told them. "That will show 'em that the Burgess is simply mistaken as to who he shot, and who shot him."

Some of them seemed annoyed with me, but the old foreman who had argued with both Major brothers that night, smiled and explained patiently enough, "No, young fellah. The time for convincing Dan's accusers of the truth is over. They've backed into a corner from where they'll put up a stubborn fight. If we tell them anything now, they'll simply adjust their testimony to it."

"He's right!" someone said.

"Then again," the foreman conceded, "we must do everything

that's consistent with our own position. Dan has been insisting that we have Bill Major arrested for shooting him. I suppose we'd better do that."

I marched with them to the home of a Justice of the Peace who signed a warrant for the arrest of Bill Major. He ordered his Constable to execute it.

On the walk back from the Justice's, I asked the foreman if he'd heard anything further from John McCann since last night.

"He came pounding at the door of Holt's at six this morning," he told me. "Drunker'n a Lord he was, an' cryin' fer shame over the harm he's brought on Dan. We didn't let him in. I had the boys take him back to his room and put him to bed. When he wakes up we'll tell him to lie low an' keep his bowels open an' his mouth shut, but sooner or later he's got to be named as the shooter."

"It's a shame he hasn't been arrested," I said.

"It is, but it's too dangerous to let them arrest McCann," said the old fellow. "If they could get the drunk to lie and say he didn't shoot the Burgess then his lies, along with George's dying declaration, would hang Dougherty fer sure."

"True", I admitted.

"Matt, there's more than enough shame here for everyone," the foreman continued. "Think of what Bill Major is going through. He knows Dan is innocent - he knows 'cause he blackjacked him to the ground. The price he's payin' in accusin' Dan to save his own skin, is lettin' his brother's murderer go free."

"That's horrible!" I said.

We went into Holt's. The doctor told us that Dan was still sleeping and didn't appear to be losing any ground.

After a while the Constable who had been sent to arrest Bill Major came in to tell us how it had gone. Bill was expecting to be arrested and he elected to treat it lightly. He handed the officer his brother's pistol and strolled with him to the Squire's office. Along the way he admitted that he had shot Dougherty but justified it as done in the due course of arresting Dan for shooting his brother.

A few minutes later there was cause for rejoicing when the doctor announced that Dan was going to pull through.

*   *   *

To this point no one involved on either side in the Mahanoy City shootings had claimed there was any involvement by any so-called Molly Maguires. There wasn't even a pretext for such a claim. But a pretext was supplied after it became known that Dan was going to survive. He was a member of the A.O.H., the Ancient Order of Hibernians, sometimes called the Hibernian Society. In return for the dues which he had faithfully paid each month – thirty-five cents – he was entitled to benefits. The benefit for dying was $50.00 for funeral expenses, and the benefit for sickness was $5.00 per week. Also, it was understood that in the event of a member being unjustly arrested or otherwise worthily in need, the Bodymaster of his local unit would take up a collection for him and, if necessary, he would ask the County Delegate to organize the raising of funds countywide. It was the fund raising efforts of the Hibernians that brought together in Mahanoy City two of the main actors in the Molly Maguire saga. I was there to take part in the encounter.

I had heard about John Kehoe, the A.O.H. Schuylkill County Delegate. He owned the Hibernian House in Girardville, which is three miles east of Ashland and eight from Locust Gap in Northumberland County. He had made a name for himself by opposing the Draft Act. He'd objected to its exemption of anyone who could pay three-hundred dollars. More recently he had used his influence in Harrisburg to get a pardon for one "Bear" Dolan, a notorious Shamokin hell-raiser. This had earned him the animosity of Father Koch, our pastor at St. Edward's parish, who knew all too well that the "Bear" belonged in a cage. Father Koch and six other anthracite-region priests had published a *Declaration Of Seven Pastors* charging that acts of murder were traceable to the A.O.H. Kehoe was instructed by the National A.O.H. to refute such charges by clamping down on lawbreakers in his local divisions.[2]

He came into Clark's on Monday morning, a lanky man in his early forties with brown hair showing wisps of grey, and when Mr. Clark introduced me he gave me a miner's hard handshake and scanned me with piercing blue eyes.

"Ah, ye're Pat McWilliams' oldest. I also know yer Uncle Tom, and yer cousin Ed who's working out of Lost Creek."

"Yes, sir, an' I've heard them speaking well about you," I said. I couldn't help liking the man and being flattered by his mention of my family even though I realized that knowing people was a politician's stock in trade.

"That's good to hear." He said sitting down at the table with me and Clark, "But we haven't yet convinced them to join the A.O.H. What's the problem, Matt?"

I was shocked by his directness. It was clear that he wanted a frank answer, so I gave it.

"They don't want to go against Father Koch," I said. "He sees your members drunk and carousing and making wild threats to break the legs of strikebreakers, and he thinks you're incapable of organizing peaceably. An' then comes the 'Bear' Dolan pardon! How would you fellows in Schuylkill like it if we released your own 'Kelly the Bum' out of jail and set him loose upon ye?!"[3]

I thought Kehoe might become angry with me telling the truth, but instead he became sad.

"I knew it!" he lowered his head and shook it. "Pushing for that pardon was a mistake! Nevertheless, it's also true that our own Bishop is an Englishman who's been weaned in his cradle on distrust for the Irish, an' he's encouraging every one of his pastors to put a stop to the Irish taking part in any organization that isn't directly controlled by the Church."

"Unfortunately, that's correct," I said. "I'm an organizer for the

---

2 Editor's Note: See: W. G. Broehl, Jr., *The Molly Maguires*
3 Editor's Note: See J. W. Coleman, *The Molly Maguire Riots*, pp. 66-67.

W.B.A. an' some of the priests have called us 'agitators' and 'demagogues'. Not Father Koch, though. He's too smart not to make distinctions. He knows the difference between a secret, oath-bound society and a labor union."

Kehoe demanded, "Then he should read the A.O.H. charter and see that our purposes are peaceful!"

"No," I said. "The only way you could convince him of that would be to act that way."

"I'll try to do it, God help me, but that is utter nonsense, Matt," he retorted. "By that standard, then, Father Koch and the Bishop are agitators and demagogues because, sure as hell, all of my bad apples belong to their Church."

I had no answer ready for that, and Kehoe excused himself to go to the outhouse.

I said to Mr. Clark, "I never thought about it quite that way."

Clark told me, "Kehoe's having trouble controlling his hotheads. His predecessor in office with the A.O.H. actually went to Shenandoah and had them elect Jim McKenna, the worst desperado in town, as their leader. Kehoe's stuck with him now. Worse'n that, his own brothers-in-law are by far the most active troublemakers between here and Girardville. He's stuck with them, also."[4]

I admitted, "The W.B.A. is hard pressed to keep the lid on its own pot. The Shamokin unions won't join us because our President is generally against going on strike."

Clark added, "You can add to the pot that the entire region is a hornet's nest of Orangemen and Know-Nothings."

"We have got to stop ourselves from thinking like that, or we'll never be united!" I exclaimed.

Clark didn't disagree but simply dismissed the idea with a shrug – of despair I thought.

Kehoe returned and efficiently processed Dan Dougherty's request for a countywide fund-raiser. He had been assured by everyone on

---

4 Editor's Note: See Broehl 262, *MJ*-Dec.17, 1875

our side of the deadline, including the priest who had heard Dougherty's confession, that Dougherty was not guilty.

"I hate paper work," he growled but he sat himself at a corner table and forced himself to grind out letters to all the A.O.H. Bodymasters in the County, instructing them to collect funds and remit the proceeds to him.

"This letter writing will take all day!" he complained.

I went to help canvass the town in search of witnesses to the firefight and the shootings. Returning to the Emerald House just as the daytime shift was let go at the mines, which was 5:30 p.m., Mr. Clark told me I was in time to join him and his wife and Mr. Kehoe for supper. Kehoe pushed aside his correspondence saying it was almost finished, and Mrs. Clark served bowls of corned beef and cabbage and some rolls. In the middle of a big mouthful of cabbage, Mr. Clark pointed toward the door.

In came a man of most remarkable appearance – of middle height, but very thin and wiry with a stealthy kind of walk like the pace of a stalking cat. He had a wide-mouthed, thin-lipped face, as drawn and cadaverous as a skull, and heavily-lidded slits for eyes, which darted around the room suspiciously. When he sat at a table and took off his hat, I was surprised. Though in his early thirties, the top of his head was completely bald and what hair remained on the back of his head and sideburns was long and lank and prematurely white .

Mr. Clark said, "That albino feller there is one of the bodyguards for the infamous Jim McKenna who I told you about earlier. After he's had time to make sure that it's safe in here, McKenna will come in."

"Ah, yes, the Bodymaster from Shenandoah," I acknowledged.

Kehoe explained, "McKenna is Secretary of the division, technically. But he had us appoint as Bodymaster a young fellow who can't read or write, who works ten hours a day, and who does whatever McKenna tells him. So he may as well be Bodymaster."

"There's Tom Hurley, another of his henchmen," Clark said as a thuggish type of fellow entered. "McKenna brought him with him from Pottsville.[5] He won't be long to appear."

McKenna walked through the door in what I later learned was his most presentable travelling condition. He had on a decent coat and trousers, and his auburn hair and beard were neatly trimmed. He was in his late twenties, of medium height and build but broad shouldered, with a wide forehead, wide-set eyes and handsome, well-proportioned features. The only thing making him worthy of more than casual notice was that under his open suit coat, in holsters whose straps crisscrossed his back, he carried two good-sized revolvers.[6] He spied us, smiled, and swaggered over in a devil-may-care manner. "Hello, Mr. Kehoe! G'day to ye Mr. and Mrs. Clark. I've just been to see Dan Dougherty, an' pleased to say it, he's feelin' good enough to be in a bad temper. Heh, heh!"

"Hello, Jim. I have a letter here for which ye can save me the postage," said Kehoe searching through his pile of stuffed envelopes. "In the meantime, I'd like you to meet Matt McWilliams, an organizer for the W.B.A."

His hand wasn't that of a workingman. But his interest seemed genuine as he asked, "Ye're a bit young fer that, aren' ye?"

"Yes, I've only just begun," I admitted.

He abruptly turned to the others and asked, "An' who was it, can ye tell me, who actually shot the Burgess?"

"All I know is that it wasn't Dan Dougherty," said Kehoe, handing him the envelope.

"We've heard different stories," said Clark. "One theory is that it was his own brother Bill, who thought he was shootin' at the boys of the Humane Fire Company. The whole trouble came about through a quarrel between them and the Citizens' Company."[7]

I was surprised that Kehoe and Clark were not telling this man what most people already knew – that John McCann was the shooter.

---

5 Pinkerton, pg. 82
6 Pinkerton, pg. 226
7 Pinkerton, pp. 232-233

I must have looked like I was about to blurt out the information because Clark nudged me under the table, which I took as a caution to be quiet.

McKenna's brow furrowed and he seemed to want to ask further questions but instead he shrugged, smiled shrewdly and said, "Doesn't that beat all! Well, I'll repair to the bar and lave ye to enjoy ye'r supper."

When he was out of earshot, I whispered to Clark and Kehoe, "Ye don't trust this McKenna, then?"

"It's not that," said Kehoe. "He's got a wild bunch up there in Shenandoah. Look at them! They want to settle everything with guns. If I should let them get involved with the Modocs down here, who are of a similar bent, there'd be hell to pay."

"An' I simply don't trust him," Clark acknowledged.

"Why not?" I asked.

"He's a fake," said Clark. "He's made a name for himself as a barroom boxer. But one night here he challenged a man to step outside. When the fellow went, old Whitey there was waiting in ambush. The man said that Whitey covered him with a pistol while McKenna beat him with brass knuckles. I didn't witness it personally, but I know for a fact that Whitey was out there, and the fellow's bruises bore him out. I believe him."

"That's nasty business!" I shuddered.

I had opportunity then to take some measure of McKenna. I could see why he'd be well-liked in a barroom. He was quick to flash a smile, and though he'd sing a song or tell a story at the drop of a hat, he didn't otherwise monopolize the conversation. He acted as though he liked everyone he met, asked many questions, and took a genuine interest in their answers. In no time at all, he learned from three different men that John McCann had shot the Chief Burgess.

He bought drinks for everyone. When I went to the bar he offered me one. I ordered a pint of beer. He seemed disappointed with my choice and joked that only whiskey was a fit drink for a man. I argued to the contrary. Then by way of rebuttal, he poured a measure for himself and launched into the old song about the full juglet, the cruiskeen lawn:

> Let the farmer praise his grounds,
> Let the huntsman praise his hounds,
> The shepherd his dew-scented lawn;
> But I, more blest than they,
> Spend each happy night and day
> With my charming little cruiskeen lawn, lawn, lawn,
> My charming little cruiskeen lawn.

He had a fine baritone voice and when he got to the chorus he waived his arms to encourage everyone to join in, which they did with a Gaelic exuberance that rocked the room.

> Gra machree ma cruiskeen,
> Slainte geal mavoureen,
> Gra machree a coolin bawn, bawn, bawn
> Gra machree a coolin bawn.

He challenged me to a game of darts. Although he was quite practiced at it, the emptying of his full cruiskeen and the mustering of all of my still-sober concentration enabled me to eke out a close win. He wasn't a very good sport about losing. He used some barroom tricks to distract me when I was throwing, and his congratulations seemed forced.

A neighbor with an accordion played "Gary Owen" and McKenna's mood brightened. He seemed in his true element as he danced in time with the fast music. He had on heavy miner's boots, and yet his legs moved so swiftly they seemed to blur. And there was nothing awkward about him. He was all poise and style and grace; clearly the student of some exquisite Irish dancing master. He finished with a flourish and a bow to thunderous applause, which I thought was well-deserved.

After catching his breath and more whiskey, he elected to sing "Brennan On The Moor." But he changed the words. He named Dougherty rather than Brennan as the highwayman who used a shotgun on the town's Burgess, and he changed the name of the town to Mahanoy City.

There were some who found his play of wit amusing, but with George Major's life draining away at that very moment and a noose being fashioned for the innocent Dougherty, most of us found the performance to be in shockingly bad taste.

Kehoe was especially upset and he drew McKenna aside and said, "Fer Christ's sake, Jim! I'm trying to convince people that we're a friendly and benevolent society. This kind of belligerence doesn't help at all!"

Embarrassed by the censure, McKenna allowed his anger to show and retorted, "It's only a bit o' fun! What the hell's wrong wid ye, Jack?!"

Hoping to take some of the heat out of the confrontation and to end the evening, I interrupted, "Sure it's been lots of fun, but now it's getting late. Sing for us 'The Parting Glass', won't ye, Jim?"

McKenna cast me a look of icy malice, but then he shrugged and said, "It's not my bedtime."

His henchman, Tom Hurley, pretended to bump into me by accident, and with an evil leer he tried to fashion a sarcastic joke saying, "If ye wants to sing, go over by the door an' I'll help ye out!"

I told the ruffian to "Bug off!" and turned away.

Kehoe grunted good night to everyone and with his leather bag of correspondence under his arm he left for his home in Girardville.

I went up to my room shortly thereafter. McKenna was casting dice with an equally tipsy merchant to see which of them would buy the next round of drinks. I heard him singing as I was falling asleep.

At that point I would have agreed with Mr. Pinkerton's summation of McKenna's character, with the exception of his assertion that anyone could drink all that whiskey and only "assume" to be a drunk. He says, at page 147:

> *His jolly - devil-may-care manner,*
> *his habit - not really a habit,*
> *but an assumption of one —*
> *of being nearly always intoxicated,*

> *ready   and   willing   to   sing,*
> *shoot,   dance,   fight,   gamble,*
> *face a man in a knock-down or a jig,*
> *stay   out   all   night,   sleep   all   day,*
> *tell   a   story,   rob   a   hen-roost*
> *or   a   traveler – just   suited   those*
> *with whom he daily came in contact.*

We are left to ourselves to speculate exactly how McKenna merely "assumed" his habit to rob a hen roost or a traveler. When I now read that passage and realize that Pinkerton is trying to say something nice about his employee, I am fortified in holding the much-worse opinion of him that I later developed.

Chapter 6

The Trial Of Dan Dougherty

A few days later, on November 6, the strangest phenomena in this strange murder case began to unfold. The newspapers began to blame the whole affair upon the Molly Maguires!

I had heard about Molly Maguires before, from people who had read about them in the newspapers. But I had never met anyone who claimed to be a Molly. I had never seen any written statement by a person claiming to be one. I knew there were some who applied the name to anyone who was Irish, much the same as they used the term "Mick" or "Modoc." But you'd think that if there was a real organization which had taken the name "Molly Maguires" (which was the newspaper's claim), then at some time someone would have overheard its members apply the name to themselves. I don't think anyone ever so testified.

If you read the testimony in the Dan Dougherty case, you will not find a single mention of the name Molly Maguire. So it was a complete shock to me when I read in the November 6 *Miners' Journal* that the shooting of the Burgess was part of a carnival of crime ringmastered by the Molly Maguires! I knew it wasn't true. It was just as false as the newspaper's pronouncement that Dougherty was clearly guilty.

George Major had died at 10 o'clock in the morning, Tuesday, November 3. The howl of rage from the anti-Irish community was as though they hadn't been expecting it. In the afternoon a physician from Pottsville, along with a reporter from the *Miners' Journal*, had been escorted by armed police into Dougherty's bedroom. Dan's friends had permitted them to observe his wounds but not to palpate, probe or treat them. In its November 6 issue, the *Miners' Journal* reported that Dougherty had been "taken red-handed" and their doctor was of the opinion that "he was only shamming" about being seriously injured.

The article asserted that George Major – portrayed not as a belligerent, liquored-up dog shooter but as a paragon of law enforcement – had been deliberately targeted by the Molly Maguires. Its recommendation for Dan Dougherty was that "one good, wholesome hanging, gently but firmly administered, will cure a great deal of bad blood, and save a great many lives in this community." This tour-de-force of investigative reporting closely aped the ravings about Molly Maguires by the *Shenandoah Herald*. I seldom read that local rag. I refer to it, primarily, to show that a fair trial for Dan Dougherty in Schuylkill County was patently impossible because of press hysteria about Molly Maguires.

The remainder of November and December 1874 saw a number of clashes between the factions in and around Mahanoy City. An attempt was made to burn down the Catholic Church because its pastor refused to publicly condemn Kehoe's A.O.H. fundraising for Dan, though the priest very vocally condemned the kinds of violence of which the Mollies were accused. Blurring the distinction between the A.O.H. and the Mollies, as usual, was McKenna's notorious henchman, Tom Hurley. He robbed a shipment of clothing and sold it to raise twenty-five dollars towards the Shenandoah Division's contribution to Dougherty's Defense Fund.[1]

In the meantime, John McCann started making his way back to Ireland, where the only drunks permitted to shoot off firearms are the Gentry.

In January of 1875, the W.B.A. was goaded by Gowen into the

---

1 Editor's Note: See Broehl, pg.182, citing *Reading Papers*

disastrous Long Strike. In some places the officers of the A.O.H., such as Kehoe, were also union leaders and they used A.O.H. contacts to coordinate union activities.[2] Therefore, they and the A.O.H., under the sobriquet of Mollies, were accused by the press of violence brought on by the strike.

Kehoe met with the State and National leaders of the A.O.H. and convinced them that the way to combat these accusations and to assert their central control over their local divisions, was to require all the A.O.H. members in the County to march openly, proudly and peaceably in the March 17 Saint Patrick's Day Parade, to be held that year in Mahanoy City. That the A.O.H. wanted to demonstrate its peaceful nature was widely publicized by word of mouth and was attacked by the press. When the day came, all four-hundred men marched precisely as Kehoe ordered them. There was not a single instance of drunkenness or brawling or disorderly or improper conduct of any kind.[3] This gave the lie to Pinkerton's claim, expressed in his doctored detective's reports,[4] that although Kehoe preached against violence, he would have to supply murderers to the Bodymasters if they demanded them. Unfortunately, peace and quiet and discipline are seldom considered reportable news for the papers. Father O'Reilly of Shenandoah, who daily had to deal with Jim McKenna, Tom Hurley, "Kelly The Bum" and their less renowned ilk, went to the parade and took down the names of the marchers from his parish. He read them from his pulpit the next Sunday, telling the congregation to pray for them as Molly Maguires and lost souls.

In addition to my job with the Union, I worked without pay as a student-clerk for the lawyers defending Dan Dougherty. His case was called for trial in Pottsville on March 28, 1875, and the jurors were empanelled. But we presented the Court with so many newspaper clippings of inflammatory invective against Dan and Molly Maguires that the judge transferred the case. It was moved to the farming town

---

2 Editor's Note: See Lavelle, pg. 277.
3 Editor's Note: See Broehl, pg. 190-191.
4 Editor's Note: See *Reading papers*, Ap. 29 and May 26, 1875.

of Lebanon, eighty passion-free miles away. There, the trial began on April 2, 1875.

It's gut wrenching for the innocent defendant in a criminal case to sit and listen as hammerblows of incriminating testimony are piled on by the prosecution. Most people want to scream "Stop!" and close up their ears. Defense attorneys know that this is the time to listen carefully.

The lawyer making the opening statement for the prosecution promised to prove that in the course of the quarrel after the fire that night, Dan Dougherty (rather than McCann) shot George Major in the breast and the deceased then pulled his revolver and shot Dougherty in the neck.

For two days, including special nighttime sessions, we listened to the prosecution's eyewitnesses and I wanted to scream at them, "You idiots! Your description of the man who shot the Burgess exactly fits that drunk, McCann! Why do you insist it was Dougherty?!"

Of course, I didn't actually scream at them, and I finally realized they were lying, though half of them were policemen, sworn to uphold the law.

None of them admitted to seeing Bill Major lunge at us and let fly with his blackjack before the shooting. None of them would say anything about how the Chief Burgess had advanced upon us while waving his big pistol. They tried to avoid telling the jury what had happened that night prior to the shooting – the water spraying, the Burgess' anger about the spraying, his killing of the dog – which would have explained the passions leading to the shooting. Most disquieting of all to one sincerely trying to understand what happened that night was Bill Major admitting that he shot off his brother's pistol shortly after it was handed to him. He would give no reason for shooting it. He said he pointed it straight ahead of him and fired, but he did not aim at any person and he saw no person in that direction. He was obviously lying.

At the end of the two days, before meeting for supper with the rest of the defense team, our lead counsel, Francis W. Hughes, invited me to his hotel suite for a consultation. Mr. Hughes had been a State Senator and Secretary of the Commonwealth and was a leader of the Democratic Party in Schuylkill County, and I felt privileged to be assisting him. Only four weeks previously, he had won praise from the Modocs and the *Miners' Journal* for having successfully defended the notorious Welsh pistolero, Gomer James, who had shot and killed a member of an Irish clan named Cosgrove. He was an effusive man, given to flamboyant gestures and dramatic statements, and a bit of an aristocrat. But he wasn't too proud to proclaim that he was for hire to anyone who could pay his ample rates. In politics, he seemed to be a genuine friend of the laboring man. He poured us both a drink, which I couldn't refuse without incurring his suspicion. Then throwing himself into an easy-chair he asked, "Well, Mr. McWilliams, what do you think of the case thus far?"

I realized he was searching for the visceral reactions of a juryman to what had been presented.

"So far, the witnesses all say that Dan shot the Chief Burgess. They're lying, but how do we prove it?"

Mr. Hughes sipped his whiskey and asked analytically, "Has their story raised any doubts?"

"A bundle of 'em," I answered. "The pistol they put in Dan's hand is too big to have shot the tiny bullet which they dug out of George Major. No one connects Dougherty with any such weapon prior to the shooting."

"Fair enough," Mr. Hughes said. "What else?"

"The witnesses who remember what the shooter was wearing on his head unanimously say it was a slouch hat. But they admit Dan normally wears a cap, and the prosecution's own doctor says he saw Dan wearing a cap before the shooting."

I offered nothing more and Mr. Hughes challenged me, "You said there are a bundle of doubts. That's only two."

"It's hard to put into words," I apologized. "But the jurors must be asking, 'Is that it?' and thinking there must be more to the story. Why the hell would Dan shoot the Burgess? Because he was telling him not to fight? Nonsense! And with a bullet through the back of his head like that, how could Dan wrench free of the hold of four policemen? Finally, what about that weird testimony from Bill Major about shooting off his brother's pistol after getting to the boardwalk? He says he shot it toward the railroad for no reason and he denies he shot at Dougherty or at anyone at that time. What about that?"

Mr. Hughes smiled and said, "That's good, Mr. McWilliams. It's good that the jury wants to hear the rest of the story."

Mr. Hughes had decided not to include me on his list of defense witnesses because I knew one thing that he preferred to keep from the jury. After the fire I'd gone with Dougherty into his hotel where McCann had shown us his pistol. Even though Dan had told McCann to put the thing away, Mr. Hughes said he did not want to volunteer any prior connection, no matter how slight, between Dan and the murder weapon. Mr. Hughes tried to convince me that I could tell the truth about that night without mentioning my seeing the pistol at that time, but I insisted it wouldn't be the "whole" truth without it.

Next morning the prosecution put on as its final witness, a house carpenter, who was boarding at Holt's that night. He saw Dougherty come in after the fire with "a stranger" (who was me) and later he heard Dougherty say to someone, "Don't let that child see the revolver!" He didn't see the pistol itself, or to whom Dougherty was talking.[5]

The prosecuting attorneys tried to act as though this testimony clinched their case, but in my opinion the truthful carpenter hurt them badly. Obviously, Dougherty didn't have possession of the pistol when he told someone to hide it from the child. Furthermore, the carpenter was the one witness who confirmed with certainty that Dougherty was wearing his cap when he led us out to roll up the hose, and was still wearing it when he returned, suffering from a shot

5 MJ, Ap. 27, 1875 test. of Thomas Kaercher. See subsequent MJ issues for the defense testimony.

in the back of his neck. Finally, the fellow practically opened up the defense case by admitting that shortly after Dan came in shot, another man came into Holt's who was also bleeding from the back part of his head. He described this man as five-foot-eight, with a small chin beard and moustache and not very attractive – a perfect description of McCann. The defense asked the carpenter whether this man complained that he had been shot in the head. The Court prevented the witness from answering that question. Everyone in the courtroom knew what the answer would have been. And this was where, in substance, the prosecution rested its case.

Our defense team then produced eight witnesses who confirmed everything the prosecution witnesses had said about the shooting of George Major except:

(i) After challenging everyone in the Humane Company to a fistfight, Bill Major had let fly in all directions with his blackjack; then

(ii) It was John McCann in a slouch hat who shot George Major with his little pepperbox pistol and whom in return had his scalp creased by a bullet from George Major.

This testimony in contradiction of the prosecution's witnesses was sufficient to establish a reasonable doubt as to the guilt of Dan Dougherty, I thought.

In addition, we produced a number of people who were in Holt's that night who saw that John McCann had suffered a gunshot wound to the scalp. Our doctor verified that the wound was, in fact, a shallow crease from a gunshot. The prosecution tried desperately to prevent the witnesses from recounting John McCann's admissions that night that he had shot George Major with his little pepperbox pistol. But McCann's landlady who cleaned and bandaged his head in her boarding house blurted out, "He showed me the revolver he shot Major with!" The court instructed the jury to disregard that admission, but there it was. As a final straw, we had our Justice of the Peace testify that McCann had come to him to have George Major arrested for shooting him. This was enough, I thought, to establish

Dan's innocence by a preponderance of the evidence.

What established Dan's innocence beyond a reasonable doubt was the ballistic evidence. We had nine witnesses who had seen from various vantage points that the bullet that entered Dan's neck was shot by Bill Major with the pistol handed to him by his dying brother. Bill's story that he had shot his pistol at no one, to the extent it was believable, confirmed these people. Our Constable testified that when he arrested Bill for shooting Dan, Bill had admitted shooting Dan.[6]

A surgeon had been called in by Dan's doctor the morning after the shooting to examine Dan's wound. On the witness stand he testified that during the night he had extracted the bullet from Dan, and he held it up to show the jury. It was a large caliber to match the pistol of the Major brothers. He said that the explosion of such a large pistol, if occurring close to Dougherty's head, would have left an indelible mark that could not be washed off, commonly called powder burns, and there had been no such burns. Dougherty must have been shot from some distance away. This testimony was confirmed by Dan's treating physician. Still another doctor and also Mrs. Holt, verified there had been no powder burns on Dan's head. Then we put on an expert who had been a surgeon in the army, and most knowledgeable about gunshot wounds. He testified there was no doubt the bullet taken from Dougherty fitted the Major brothers' pistol, but in the absence of powder burns there was also no doubt that it had not been fired at close range by one holding Dougherty by the coat.

The Judge explained to the jury the significance of this testimony, pointing out that the Commonwealth was basing its accusations against Dougherty on the claim that George Major immediately shot

---

6 Editor's Note: The prosecution was ready with twelve witnesses to testify that they would not believe the Constable on his oath. It was not effective impeachment. Nine men were found in the courtroom that testified otherwise. One of the prosecution's impeachment witnesses was Frank Wienrich, a butcher in Mahanoy City, who was subsequently accused of being a member of the vigilante squad who committed the Wiggans' Patch murders.

the man who had shot him. He told them that if they found that George Major did not shoot the prisoner, but rather the prisoner was shot by Bill Major, they must issue a verdict of not guilty.

In less than two hours, at 11:00 a.m., the jury did return with a verdict of "not guilty" and the couple of hundred or so supporters of Dougherty who had come down to Lebanon on the train broke into raucous cheers, with whistling and stomping of feet. To restore order, the Judge threatened them with arrest for contempt of Court.

Never, I think, was a crowd less in contempt. They were cheering the Court and the jury and the manifestly just result. I felt elated along with them, and proud to have played my part. My admiration and profound respect for the intricate wisdom of the law, and its codification of human experience through centuries of patiently recorded precedent, was welling in my breast. I wanted to continue to be a part of it. I wanted to be a lawyer.

Before he boarded the early afternoon train back to Pottsville, Mr. Hughes and I were enjoying one of his generous measures of whiskey and I asked him, "How could all those prosecution witnesses take an oath to tell the truth, and then get on the stand and lie like that?"[7]

"Aren't you being a bit too judgmental about them?" Mr. Hughes proposed. "Most of them probably thought they were telling the truth, or perhaps only shading it somewhat."

"No," I argued. "The fellow who said he shoved his lamp right in the shooter's face and identified Dougherty was lying outright! And so was Bill Major!"

"True enough," Hughes agreed. "Those two, at least, were the conspirators. A conspiracy to lie is a damnable thing. It has power because it gains the support of a lot of people who are its unwitting dupes. . . . who are its true believers."

"I shudder when I think that Bill Major made an unwitting dupe

---

7 Author's Note: When he learned of the verdict, McKenna's reaction was to think " . . . there will be a great deal of trouble, as all the witnesses for the Commonwealth committed perjury, and knew they were doing so . . ." P. Agency summary dated May 2, 1857.

of his own dying brother!" I said with revulsion.

"Yes," Mr. Hughes said reflectively, "George thought he was telling the truth when he said Dougherty had shot him. Then he went on to tell the whole truth about shooting back, and the truth set Dougherty free."

Clearly, the Modocs of Mahanoy City, enabled by the newspapers, had tried to hang Dan as a Molly Maguire even though many of them knew he was innocent. When he returned home someone tried to reverse the effect of the jury's verdict by taking a shot at him in the middle of town in broad daylight. No one would identify the shooter. The deadline in Mahanoy City had been rendered more deadly than ever.

In his lair in Philadelphia, Franklin B. Gowen read the daily transcript of the Dougherty trial as published in the *Miners' Journal*. His interest in the case as a Molly outrage was more than casual. His Pinkerton operatives must surely have told him what McKenna had learned at the very beginning – that McCann was the actual shooter. Gowen knew Dougherty was innocent.

Yet Gowen never interfered with the *Miners' Journal's* stubborn insistence that Dougherty had murdered the Burgess on behalf of the Molly Maguires. He could have interfered. That paper was his mouthpiece in the coal region. He maintained expensive daily advertising on its front page.

It is reasonable to conclude that he was, at least, complicit in the newspaper's ravings about the Molly Maguires. I suspect he encouraged them. He would soon viciously use the lie about Dan and Molly Maguires to convict men of murder in trials that he personally would prosecute.

Chapter 7

The Gowen / Pinkerton Arrangement

From Lebanon I took the cars directly to the Mount Airy station in Philadelphia to have my meeting with the famous Franklin B. Gowen. I had written to him that I was finished reading all the books he had sent me, as well as almost all on his list, and he replied by telegraph:

"Am delighted with your decision to assist with the Dougherty case, an experience more valuable than the books. If possible, report here in person when verdict in. - F. B. Gowen."

It remained daylight for most of the ride, and I took pleasure in watching the neatly fertile Dutch farms roll past the window. They had a stolid, functional honesty about them. They helped me to shrug aside my growing premonition that honesty and truth were becoming lost in the press of events.

While I'd been engrossed in Dougherty's struggle not to be hanged, the Long Strike had entered its fifth month. Things looked bad for the W.B.A. in Schuylkill, Luzerne and Carbon Counties and for the independently striking unions around Shamokin. The mines in the Northern field were still at work, busily producing more anthracite than ever before, while the nation-wide depression had reduced demand for it in the rolling mills, blast furnaces and iron works. Few, if any, coal operators were making a profit. None who shipped on the Reading dared to defy Gowen's order that the pay of the workers must be cut by twenty percent. The economic downturn

had reduced the prices of food, and other consumer goods, as well as
wages in other industries, and it was the consensus of the press that
it was only fair for the miners to bear their share of hard times.[1]

Two months into the strike the pockets of the mineworkers were
dry of cash. In another month some of the storekeepers who were
accustomed to supplying them with food and provisions "on tick"[2]
began to go out of business, and there was real hunger in many
households. The W.B.A. began distributing flour. Violence surfaced.
A railroad car on its way to Ashland was diverted by error to Locust
Gap where it was broken open and ten barrels of flour stolen. On
March 29 mobs assembled at several collieries in the Hazleton area
and disarmed the private police stationed to protect them. The men
working the collieries in violation of the strike – called "blacklegs" or
"scabs" – were forced to join a march to other mines and into the city
of Hazleton. Some of the policemen were beaten. One was shot in the
head, but not fatally. The marchers made speeches and forced one
operator to sign an agreement. Then, at the behest of Father O'Hara,
they disbanded. A similar march took place around Audenried, with
the same result at the behest of Father Warren.

The *Miners' Journal* reported all these strike-related events along
with the details of some suspicious deaths, some robberies and
unrelated crimes, and it proclaimed with blatant demagoguery:

> "This state of things constantly growing worse has been in
> progress for years . . . It involves the coal district generally . . . The
> cause is to be sought in a chronic state of strike which reigns
> among the miners . . .
>
> "Last week's crimes exceed anything on record. The Eckley
> mines are now guarded by a strong force . . . Vigilance Committees
> are being formed . . . The Pottsville Militia has been ordered to
> take the field at a moment's notice . . . It is evident that for the scale
> of their operations, the Molly Maguires are a more dangerous,
> desperate and audacious gang of bandits than the Klu-Klux,

---

1 *NY Tribune* reported in *MJ*-4-30-75
2 Author's Note: i.e.-on credit

and no half measures will be sufficient to put a quietus on their vandalism."[3]

Placing the blame for everything upon the Irish under the name Mollies was galling in its audacity and unfairness. There was no one to challenge it.

In the Shamokin area the strike was both effective and peaceable, for a time. In fact, the Shamokin correspondent of the *Miners' Journal* was fired for saying so. The streets, stores and saloons were more crowded than usual with the idle men. Everyone talked about how soon the operators might be forced to resume work on the basis agreed upon in '84. Then blacklegs were brought in from other Counties and four collieries were opened half the time, and soon four more. Yet over many weeks, including all through March, there was no report of violence or threats of violence against those who chose to work at the reduced rates. This was in the face of the statistical fact that in Coal Township and Mt. Carmel a very large number of orders on the poor funds were being issued to relieve want. Trade was reported as being very dull.[4]

The pressure exploded on Wednesday, the last day of March. In the dark of early morning, the boys got two coal cars, one properly loaded and one filled with combustibles, and coupled them together and set them on fire and shoved them down an incline to crash in bright, pyrotechnic splendor. They say you could see it from town.

Shortly after noon, a hundred or more strikers jumped aboard a coal train and rode it through town, hooting and yelling at everyone they saw. The correspondent of the *Miners' Journal* complained, somewhat prissily I think, that the boys rode without paying the fee.

But more seriously, the men had got word that the Sheriff of Northumberland County was on his way that day to evict a number of the striking miners from their homes in Mt. Carmel. Their leases had a clause allowing the coal company to evict them should they go

---

3 MJ-3-30-75
4 MJ-4-23-75; 5-6-75; 5-8-75; 5-12-75

on strike. Only a public display could save them. The hundred or so
who commandeered the train were joined by others from Locust Gap
and Centralia. Three hundred of them assembled with a drum corps
and marched, accompanied by martial music, to intercept the officer
of the law. When the sheriff exercised sound judgment and did not
show up, they went home.

So far as I know there was no violence against any person in the
Shamokin area as a result of the strike until Wednesday, April 21. A
wagoneer from one of the Independence Street dry goods stores was
using his vehicle to transport blacklegs to work at the Brady Colliery.
That's a quarter mile from our home on Tioga Street in Coal
Township. The teamster and his human cargo were almost at the
mine when their way was blocked by the women of the Brady and
Greenback patches. These mothers and sisters of hungry children
were in deadly earnest. They lifted a side of the wagon and tipped the
men out of it. Had the blacklegs not run back to town, they would
have been injured. The teamster and his empty wagon were reported
as "flying" past the Henry Clay a few minutes later. Clearly, these
women had run out of patience. They were not only in defiance of
the law, but in defiance of their men whose union was holding them
from taking such action themselves. It did not bode well for peace in
the future.

I stepped off the train at the Reading Railroad's Mount Airy station
and went to the freight delivery dock to pick up my crate of Mr.
Gowen's books. Several people were waiting to receive their property.
As I went to join them I spied my box. It was being loaded into the
back seat of a surrey by a large man in a slouch hat.

"Hey!" I called to him.

He put down the box and turned to me and in the light of the gas
lamps I had my first look at his face. It so surprised me that my own
face must have contorted. He was the first Negro person I had ever
seen. He looked at my wide-eyed amazement and he understood the
reason for it, and it amused him. His dark features cracked into a
dazzling smile and he said, "Mr. Gowen sent me for you."

"How did ye know when I'd get here?" I blurted.

"Sir, Mr. Gowen is President of the Railroad. His ticket agent telegraphed," the man patiently explained, still amused.

There was room for me in the back of the surrey but I hopped onto the seat next to the driver, put out my hand and said, "Hello, I'm Matt."

"I'm Gilbert," he took my hand. He was in his mid-forties with a full but short-clipped beard and slightly greying temples, six inches shorter than me, but broadly built.

He flicked the reigns and drove the horse slowly through a very fashionable neighborhood of suburban estates with landscaped gardens, well lit with gas lamps.

"I've never been in a place like this. It's beautiful," I said.

"The Gowen Estate is called 'Cresheim'," said Gilbert. "It's only four blocks away, but Mr. Gowen said you were returning books, so I should drive."

"That's good," I said, and told him, "Ye're the first Negro I've ever met."

He laughed and said, "I could tell by the look on your face."

I laughed with him, and as we passed under one of the gaslights I noted, "Ye're dark. But not nearly as dark as me when I'd get out o' the mines. No one could tell us apart down there."

Gilbert looked at me and said, "I buried both black and white outside of Richmond after the Battle of the Crater. They all had the same color blood."

The Estate was on twenty acres enclosed by a tall wrought iron fence, with graceful expanses of lawn and planted with every kind of tree and shrub. The lilacs were in bloom, reminding me of the Walt Whitman poem and spreading their sweet fragrance throughout.

The house itself was huge, entirely of brick, featuring a fancy, concave mansard roof. Gilbert promised to take care of the books and my travelling valise and ushered me into an office or "study" where Mr. Gowen greeted me.

"Matthew McWilliams! When last I saw you, you were a babe in arms. It's a pleasure to meet you in person, my boy!"

He was a handsome man. Of average height and build, he had a high forehead with wide-set eyes, a strong nose, square chin, and wavy brown hair. His smile was warm and sincere and your first impulse was to return the goodwill that he exuded.

He said that he and his wife and daughter had delayed supper until my arrival and he escorted me into the dining room to meet them.

The daughter, named Esther after her mother, was a very pretty schoolgirl of about fourteen years, well spoken but quite shy and reticent. Mrs. Gowen was elegantly beautiful and, to me, most charming. She was so full of questions about my mother and father and sister and brothers that I scarcely had a chance to eat. She asked what was happening in Sunbury, the County Seat of Northumberland County and her own home town, and she seemed mildly disappointed when I had to admit that I seldom went there because my duties for the W.B.A. kept me mostly in the Shamokin region.

The supper of roast beef, mashed potatoes, vegetables and a delicious crème brulee for dessert was a very enjoyable experience for the most part, but I noticed there was a kind of tension, a discord, between Mr. Gowen and his family.[5]

He joked with me that he might try to pump out of me some confidential information about union affairs, and Mrs. Gowen became visibly angry with him and slammed down her fork.

I quickly tried to lighten the mood by saying the truth – that Mr. Gowen probably knew more about union affairs than I did. She smiled at me gratefully and resumed eating.

Other times, Mr. Gowen would pass some observation or comment which seemed innocuous enough to me, but Mrs. Gowen and her daughter would look at each other and roll their eyes in disapproval.

---

5 Editor's Note: See Schlegel, pg 274

He mentioned with enthusiasm that they had been invited to a formal dinner and dance, and Mrs. Gowen scowled at him angrily as though embarrassed by the invitation. She hastened to tell me that she preferred small, informal suppers to the gala social events indulged in by the rest of Main Line Philadelphia. Then she went on to add gratuitously that her husband's fondness for great events was compatible with his proclivity to be exclusively and obsessively involved in matters of business.

Mr. Gowen was embarrassed by this but he tried to make light of it, apologizing that his excessive concern for the support of his family was a natural fault, and therefore very difficult to overcome.

After the meal Mr. Gowen took me into his study for brandy and cigars.

Seated in one of the easy chairs flanking a wood-burning fireplace, he relaxed completely and seemed to enjoy himself.

"I hope I didn't offend you with my joke about pumping you for union news," he apologized.

"Of course not," I assured him. "We want every single one of your employees to join the W.B.A., and we operate on the assumption that some of them report back to you!"

He laughed, as did I, and I added, "We have nothing to hide."

"But that isn't true about some of the independent unions, is it?" he said rhetorically.

"Perhaps not," I said in refusal to argue the issue.

He was familiar with the *Miners' Journal's* reports of the testimony and defense arguments in the Dougherty case, to date. He listened intensely to my summary of what would be in tomorrow's issue – the closing argument of Prosecutor Lin Bartholemew and the Judge's charge to the jury.

"I laughed," he said, "when I read Mr. Hughes' argument, for the defense, that the only way the shooter of George Major could have broken loose from the grasp of the four burly policemen who had hold of him, was that he had an accomplice to escape – OLD RYE!"

I laughed and agreed, "That was funny."

"Hughes made much of the fact that Dougherty had no known motive for harming the Chief Burgess," he noted and asked, "Did Mr. Bartholemew address that?"

"He did," I answered. "He said that Dougherty and Major belonged to fire companies which were feuding with each other, and in the coal regions where towns spring up like magic and contain a conglomerate of all kinds and nationalities of people with different customs and manners, life is not held so dear, and such a motive is enough for murder."[6]

"Interesting! Extremely interesting!" Gowen exclaimed, and puffing mightily on his cigar he wanted to know, "Did the jury accept that, do you think? How did they react to that?"

I thought a moment and said, "They listened to it, I think." If it hadn't been clear that Dan wasn't the shooter they might have accepted that as his motive."

"Very interesting! Very interesting!" Gowen repeated.

He continued to ask about the Dougherty case. I responded to him freely, except I never revealed that I had been there to witness some of the events myself. I didn't want to disclose the sensitive nature of my dispute with Mr. Hughes about testifying. That, I thought, was a confidential matter between me and Mr. Hughes.

Finally, Mr. Gowen could see that I was becoming fatigued after a long day in which I'd completed a trial and a journey, and he had Gilbert take me to my room.

In the morning after a late breakfast, Mr. Gowen apologized that I would have to excuse his unavailability until dinnertime[7] because he had business visitors scheduled to arrive.

"The gentlemen do not want it generally known that they are in Philadelphia," he explained to me, "so I would appreciate it if you would stay in the back garden, or your room, or anywhere in the house where you won't run into them, until I can be rid of them."

---

6 *MJ* – 5-3-75, Mr. Bartholomew
7 Editors Note: i.e.-lunchtime

"Certainly Mr. Gowen," I assured him.

"Gilbert has taken the surrey to pick them up. He should be returning shortly," he advised.

I went up to my room to get a book, my intention being to spend a pleasant hour or two reading in the garden. On the way down, I took the rear stairwell to avoid the visitors. At the door that opened out to the back porch, I noticed that Gilbert had left the gas lamps lit in the basement. The basement intrigued me because the Gowen Mansion was the most modern house I had ever been in, and I was curious to compare its hot water heating system with my da's ingenious system at home. Gowen's basement was so big that it had a separate nine-by-twelve room for a coal bin, and the loading chute went through the outside wall of the house. His furnace was enormous.

The place had several wood-burning fireplaces as well, and I saw that Gilbert had been cleaning out the bins in the basement into which the wood ashes were pushed through vents in the fireplaces upstairs.

The vent from the study had been left open and I could hear Mr. Gowen's voice almost as clearly as when I was sitting across from him.

"I've been paying you to live in the coal regions since October of '83. At first it was assumed you'd find work as a miner, but that didn't work out. Why not?"

"Ach, Mr. Gowen," said a man's voice that I had heard before but couldn't then identify, "I tried working down there, but after two days I couldn't take it. It's no kind of life. Not fer me."

"You've been living pretty high off the hog for a year-and-a-half and I've precious little to show for it," Gowen complained.

"I've given ye all the names of all the leaders of the A.O.H.," argued the voice.

"I could have had those names by simply watching the Saint Patrick's Day parade," countered Gowen.

"Your money and Mac's time have both been well spent, Mr.

Gowen," declared a third man with an authoritative ring to his voice. "Mac couldn't have accomplished what he has if he'd worked ten hours a day in the mines. He's shown a high degree of skill and a lot of ingenuity in gaining the confidence of these gangs of cutthroats. Yes, it has taken some time, but now he's done it. They consider him not only one of themselves, but one of their top leaders. They keep him fully posted in regard to every move they make."[8]

"Very well, Mr. Pinkerton. I accept that," said Gowen. "But only days ago I approved of your setting up the flying squads of police headed by Captain Linden. That's even more expense. I want it to be fully understood what I expect to get for my money, from everyone. Linden's part is easy – to protect Railroad lives and property and to arrest those acting against them.

"Mac's job is twofold. He's to point Captain Linden to the right places at the right times whenever he can. But more important than that, and harder to do, Mac has got to build me good cases against the Mollies. Cases that I can prosecute to convictions. I don't want any embarrassments like that Dougherty fiasco.

"By this time next year I expect you to have under arrest at least one top Molly from each of the coal field Counties."

"That's a tall order!" exclaimed the man named Mac.

"Perhaps, but that might change," said Gowen. "The strike has been going on four full months. There's every sign that many of the workers and their families are at their wit's end. Violence is surfacing everywhere."

"That's true," admitted Mac.

"I brought you here," Gowen told him, "to make it clear that we need to arrest not only those who actually commit the violence, but also those who send them out to do it – the planners and the organizers. Those are the boys we want the most."

"Mr. Gowen has recently advised us, Mac," said Pinkerton pedantically, "that when our flying squads do catch the culprits in the

---

8 See the letter of Pinkerton Superintendent Franklin to Gowen. 3-13-75

act as the result of an alert from you, that doesn't mean we have to reveal your identity. And we certainly won't."

"We can prosecute the actual perpetrators for what we catch them doing," confirmed Gowen. "Then later, much later if need be, we can prosecute the ringleaders, with your testimony, of course.[9] Do you understand that? It's important."

"Aye sir, dat's clear enough. But as I tol' Mr. Pinkerton, I hope ye'll forgive me if I should have a few reservations about it."

"You do? Why so?" Gowen asked.

"If I'm in on the plannin', won't I be as subject to prosecution as the rest of 'em?"

"Of course not!" Gowen exclaimed, "No more than Mr. Pinkerton or myself! We can all prove that we're not actually aiding these villains, we're trying to arrest them!"

Mac protested, "But what the divil do I answer when they ask me why I did nothing to stop them?"

"It's a sad fact of undercover detective work," Pinkerton told him, "that an operative can't be expected to prevent every depredation committed by the gang he infiltrates."

"And the bigger and more widespread the gang, the more is that the case," Gowen added, and he advised, "Just remember the simple rule, that the protection of your own life is always a sufficient reason for not acting."

"Also remember, Mac," said Pinkerton, "that Mr. Gowen has already spent far too much money in this effort to justify us in revealing who you are prematurely. That would be a terrible waste."

"I want to convict the ringleaders and I want to get all of them," Gowen emphasized.

"Your job," instructed Pinkerton, "is to report what the Mollies are up to as soon as reasonably possible, and to let me and Mr. Gowen decide what, if anything, should be done about it. That's our responsibility."

---

9 Author's Note – See Appendix

"I always understood that the decision of when to reveal my true identity would be up to you, sir," Mac told Pinkerton. "I won't do it without your say so."

"That's a rule to which there can be no exception," said Pinkerton.

"That's settled then," said Gowen. "It isn't possible to be completely specific about how you should go about achieving the result we want, Mac. But as District Attorney, I learned that one of the best ways to make a case is to get one of the accomplices to testify against the others. And that doesn't have to be you, Mac; it can be somebody else. Understand? If you put a man's head in a noose you can usually get him to talk. Don't forget that."

There was a short silence after which the man called Mac spoke, "That's an idea . . . worth thinking about."

There was a longer silence and Mac spoke again, "Mr. Gowen, are ye tellin' me that if I find two men doing something criminal, I'm permitted to promise one of them that if he informs on the other, you'll let him go free? Scot-free?"

"Yes, I would." answered Gowen. "Though I'd expect you to make a better bargain than that – say, one man going free and a whole lot convicted."

Pinkerton hastened to say, "It depends upon the circumstances, of course."

"Ach, sir! Sure, I understand that it will all have to be on a case by case basis," said Mac. "But that's very helpful, sir. Sure, an' it is."

"That's good," said Gowen. "This entire effort must be brought to a conclusion one way or another, and fairly soon. I wanted you to know my mind about these matters. But a note of caution. Since I'm going to be one of the prosecutors, I don't want to be personally involved in the investigations themselves. I'd prefer if this meeting never took place.[10] The next time we meet, let us hope that it's to prepare the actual cases against the Molly Maguires."

As the men concluded their meeting and said farewell, the man

---

10 Author's Note – See Appendix

called Mac made a comment that sent a shock of recognition through me. He said, "I'll repair back to Shenandoah on the next train."

I then remembered I had heard that same voice using almost that same phraseology the evening Jim McKenna had told me and John Kehoe and the elder Clarks that he'd "repair back to the bar" and leave us to enjoy our supper.

"My God!" I sucked in my breath to avoid exclaiming aloud, "Pinkerton's undercover detective is none other than the notorious Shenandoah roughneck, Jim McKenna!"

My first thought was that I was probably mistaken and my second was that no one would believe it. I had to verify it for myself or, I knew, it would drive me mad.

I strode out of the basement the way I had entered, dropped my book on the back porch, and ran to the rear of the property. The iron fence was six-feet-high and spiked, but I scrambled up and over it with ease and raced the four blocks to the railroad station.

There, I stood beside a pillar next to some waiting passengers and watched for the Gowen surrey to arrive, which it did shortly. The curtains were down, hiding the occupants, but they got out close to me.

First was a fifty-or-so, redheaded gent who was huge, perhaps even taller than me, and dressed as a bit of a dude. Pinkerton, I assumed.

Then bounded out with irrepressible energy a freshly cleaned, pressed and sober Jim McKenna!

"I've seen it with my own eyes!" I said to myself and slowly walked back to Gowen's house.

I knew that I had no idea what I had gotten myself into, and needed time and more information – perhaps a lot more – to make a rational decision what to do.

It took all of my will power to drive the incident from my mind and to complete my visit with Mr. Gowen as though it hadn't happened. My experience of staving off my fear of working in the mines helped me to do it. We had a pleasant afternoon. Mr. Gowen pronounced me capable of being a credit to the bar, and he put me on the evening train with assurances that he would recommend me to read with any lawyer of my choosing. In those days, that was as much a guarantee of success as a young man could have.

So the start of my ride home was filled with thoughts of pleasant prospects. But soon enough the ridges of the Appalachians loomed out of the dusk, as did my thoughts of what to do with what I knew.

Looking back with hindsight, it's shameful to admit doing what I did. It turned out to be the wrong thing, and I'll not seek to excuse my guilt for it, but simply state how it was.

I was stunned and confused by overhearing Gowen, Pinkerton and the detective talking among themselves the way they had about the Molly Maguires – as a very real and existing organization of which McKenna was an actual member. (I wasn't yet aware that, although their correspondence shows they knew better, they were accustomed to calling anyone who was in the A.O.H. a Molly Maguire). Sure, I had never met anyone who claimed to be a Molly or who claimed to know one, and no coffin notice or threat or claim of responsibility for violence had ever been signed by anyone saying he was a Molly, and it was my conception that the so-called Molly Maguires were a mythical creation of newspaper hysteria. But that conception was now seriously challenged by what I had heard. Pinkerton had caused the highly intelligent Mr. Gowen to pay good money to maintain a detective among the highest ranks of the Molly Maguires, and that was powerful proof that they did exist. (I was too naïve to understand that conspirators to a lie sometimes maintain the lie among themselves.) And if there was such a gang of criminals, be they Irish or otherwise, I was in favor of what Pinkerton and Gowen seemed to be doing to eradicate it.

Also giving credence to what I'd overheard was the vicious nature of the gangs in and around Shenandoah and Mahanoy City. On the Irish side were the likes of Jim McKenna, Tom Hurley, Mike Doyle and Kelly The Bum, while on the Modoc side were Bill Major, the pistollero Gomer James and Bully Bill Thomas and their ilk. They were always in the newspaper for brawling or shooting or for suspicion of robbery or mayhem of some kind.[11] If there was a Molly Maguire organization, those were the places it would most likely be found, and those men were its most likely members, and their antagonists. All of them would have better served their communities from behind bars. And any one of them would have gleefully cut McKenna's throat upon learning he was a detective, I thought. I was soon to become more experienced in the ways of treachery.

Those were the reasons I decided, sad to say it, that I would keep the Gowen/Pinkerton secret to myself until I learned more about who it was likely to affect, and how.

---

11 Editor's Note - Appendix

CHAPTER 8

THE PINKERTONS' WILD GOOSE CHASE

By the time I walked from the train station to our house on Tioga Street the night of May 3, it was late and the family was asleep. I let myself in through the kitchen door, at street level, and made it up the three flights of winding stairs to my room in the attic without waking anyone. I had resolved that I'd find a quiet moment to sit them all down, tell them what I had overheard, and ask them what they thought about keeping secret the identity of Gowen's detective. I had no idea how difficult it was going to be to find a quiet moment in the week that followed.

The next day delegates of strikers from the Hickory Ridge and the Lancaster met on our back porch to decide what they could do to stop the operators from continuing to work those mines with their imported blacklegs. Both places were within a mile of the house. There was talk of setting fire to the breakers. This was countered with the argument that the process of rebuilding would delay underground employment even after the strike. It was suggested that the blacklegs be chased away with threats and, if necessary, with beatings. Arguments against causing harm to people were quickly voiced. Some were in favor of accepting the twenty percent cut in pay and returning to work. They were told that all of the suffering they had already endured would be rendered in vain by such a surrender. There was no satisfactory solution, but only degrees of dissatisfaction, for the miners. They agreed to meet again on Saturday.

That evening my da's new laborer, Dutch Henry, lingered with the family on the back porch until past sundown. He seemed to be paying a lot of attention to my sister, Mary, who, I noticed, had become a dazzlingly beautiful young woman. The mountain on which our house was situated continued to rise steeply behind it for more than a quarter of a mile, and yet over its summit to the southeast we noticed a crimson glow in the sky. As we continued to watch it, the red aura grew alarmingly higher and brighter. Dutch Henry invited our Mary to ride east with him in his buggy to learn whether there was a forest fire or other kind of conflagration. They went only three miles before they met men from the hamlet of Helfenstein, five miles over our mountain. They said they had taken up clubs that afternoon and had driven off the blacklegs at the Ben Franklin and the Helfenstein collieries, and they had returned that night with torches to burn down both of the breakers.[1]

The news was received by us with somber regret that a wage dispute had come to such a pass. It was no win for us. The operators of those breakers wouldn't win either, regardless of the outcome of the strike. Two of the operators were the same Baumgartner Brothers who, years before, had brought young Frank Gowen to Shamokin to operate their Iron Furnace. Da' said they were still associated in business with Gowen and they would certainly complain to him.

The next afternoon some railroad workers came in with news from the mine at Gordon Plane, two miles south of Ashland. There, the operators had brought in German and Mennonite farm hands from Reading to continue production in defiance of the strike. To scare them back to Reading, the miners had gone out early in the morning and applied an axe to the wire rope on which they were to be lowered down for the first shift. While the empty cars were being brought up for them, the rope parted and there was a hell of a smash of wood and metal at the bottom of the shaft. No one was hurt, but a lot of those fellows took the hint and went back to farming.

1 Editor's Note: See *Shamokin Citizen*, Centennial Issue 1964.

Nevertheless, the mine stayed open.[2]

On the evening of May 6, I thought there might be opportunity for a talk with my family in private, but again, Dutch Henry showed up. He had found part-time work in fixing the plumbing at the Ashland House Hotel, and he brought news from there. He told us that a union organizer fresh off the train was spreading a warning to workingmen all along the line. The warning was that a gang of seven ruffians had just been sworn in as Coal & Iron Police in the Pottsville Court House. Their leader was a tall, black-bearded tough, calling himself Captain Linden. They were on their way to Ashland and should probably be arriving in the morning.[3]

I remembered Mr. Gowen saying that a Captain Linden and his "flying squads" of Coal & Iron Police were all Pinkerton detectives working with the undercover operative McKenna. The affair had moved uncomfortably close to home – ten miles actually – and smack in the laps of our cousins in Locust Gap and Ashland. I had to warn them.

When I arrived at my cousin Patrick's house in Ashland the next morning, he told me that Linden and two of his men had arrived on the early train, and the others would probably follow later in the day. He had already been warned about them. In fact, he had a bellboy at the Ashland House who, for a price, was keeping him informed. Linden and one man were boarding there while the other five men had booked rooms across the street at the Mansion House. Both hotels were crowded with blacklegs. I sought out this Captain Linden to see what he looked like. He wasn't a man who could easily escape notice, being tall and burly with black curly hair and beard and wearing a pale grey, three-piece suit of the latest Chicago fashion, and a ten-gallon hat. He'd hired a local man to show him around like a sightseer and he asked questions of everyone he met. He might as well have worn a sign on his lapel saying "Pinkerton Detective!"

I was invited by my Uncle Tom and my cousins in Locust Gap to

---

2 *MJ*-May 6,1875
3 Author's Note: Franklin's summary of Linden's field report for May 6 says the new police were noticed and followed by an "Irishmaner"

have supper with them, and of course I accepted. The place was half way between Ashland and Springfield.[4] We were all sitting outside on the stoop after the meal, smoking pipes and cigars and enjoying the sweet evening air, when there was a loud explosion that reverberated and rolled off the sides of the valley.

"What could that be?!" I exclaimed.

Uncle Tom was not surprised. He smiled ruefully and explained, "Some of the boys from Hester's Barroom said they might welcome the Coal & Iron Police to the area by blowing up a bridge. I told them not to do it. They're railroad workers who were fired for joining the union. They don't know a damned thing about explosives, and I warned them that dualin isn't the same as gunpowder. They might have blown themselves up."

"Perhaps we should go and have a look," I suggested.

When we got to the place, called Locust Gap Junction, the small bridge was still standing. Railroad employees had formed a bucket line and were sloshing it with water from the creek. Uncle Tom examined the damage. Auger holes had been bored in the timbers and dualin put in and fired. The explosion had shattered some of the timbers and set them ablaze.

The railroad foreman said he'd taken the precaution of rerouting the evening train from Pottsville over a different route into Shamokin, and he'd have the timbers replaced by Saturday.

Uncle Tom spat a mouthful of tobacco juice into the creek and pronounced his one word verdict upon the efforts of the incendiaries, "Amateurs!"[5]

The meeting scheduled for Saturday was not held on our back porch because men from the Locust Gap, Excelsior, and Greenback collieries wanted to join us and the porch wasn't big enough. We met over the hill in Greenback, and it wasn't pleasant. There was a rumor that the Baumgartner Brothers were going to bring their blacklegs from their collieries that had been burned down and give them jobs

---

4 Author's Note: This is technically named Marshallton, but no one calls it that.
5 *MJ*-May 10,1875.

at the Excelsior colliery and elsewhere. And worse, they were going to make the blacklegs' positions permanent! This would have been the same as serving eviction and expulsion notices upon entire communities. My da' and Uncle Tom argued that, probably, the rumor wasn't true and the men shouldn't act as though it was true until they found out for certain. But the majority wanted to put an immediate halt to the operators' use of strikebreakers and they would agree to no delay except for observation of the Sabbath.

At exactly one minute past midnight on Monday morning they began spilling coal oil on the floors of the Marge Franklin and the Enterprise breakers near Excelsior, which is little more than a mile east of Springfield. By one in the morning both of these enormous and expensive wooden structures were engulfed in flames. The conflagration could be seen for miles. At the Ashland house, eight miles east, the hotel keeper had told Captain Linden that he sensed something was afoot, that the people's sympathies were with the miners, and he believed the workingmen would burn down every colliery. Linden saw the flames spreading across the night sky and the summary of his report for Mr. Gowen's information states:

"There was fire in the direction of Centralia, about one o'clock in the morning, but Operative could learn nothing concerning it."[6]

Gowen would soon learn a mouthful more concerning it. Both mines belonged to his railroad. They had been owned by his old mentors, the Baumgartner Brothers, but in 1871 both had been purchased by the Reading Coal & Iron Co. and leased back to the Baumgartners as operators. They would tell him all about it. The burnings would be regarded by him as a directly delivered personal insult, and as a challenge to "Go to hell!" to his Coal & Iron Police.

When the sun rose that day, things didn't get better. The boys, about three-hundred of them, hid in the woods above the Hickory Ridge mine. The early whistle blew. Surface mules were brought out of their stables, and the blacklegs from their lodgings in town began

6 See summary for Captain Linden for May 9, 1875.

to arrive. Then the strikers rushed down upon the colliery. Some had clubs and a few had guns. They surrounded the entrance to the mine and the engine houses and pump houses. They treated the arriving strikebreakers to a dose of threats and curses and denied them access, but there were no serious injuries. It was a standoff. No work could be done, and eventually the blacklegs went back to town.

Then the miners marched on to the Lancaster No. 2 near the patch at Coal Run. They interrupted work there by threatening to burn down the breaker over the heads of the workers below. But the operators had been warned what was afoot and had sent to town for a deputy sheriff who soon arrived with a few men. He was vastly outnumbered but he convinced the strikers that he was not afraid of them. He coolly insisted that he would enforce the law. He read aloud his authority to order them to disperse. He told them that if they refused to leave he would go back to town and organize a posse of vigilantes, or he'd get a troop of militia to attack them with cannons that very day.

The miners knew that the officer was legally correct. They knew he and his men could identify them if criminal charges were to be pressed. Many of them had recently fought against a rebellion. They well knew to what bloody level of violence they must descend if they were to continue their effort to close the mine. It wasn't acceptable to them. They were decent people who believed in right and wrong.

Their leader burst forth with an obscenity in his native German, and then he pointed an admonishing finger at some blacklegs who were milling about and he growled in English.

"De sons-uf-bitches! Let 'em work, then, if dey are mean enough to do so!"

He waved his arm at his striking miners, a gesture of frustration and retreat, and one by one at first, and then en masse, they turned away and melted back into the surrounding woods. It was the beginning of the end of the strike in our area.[7]

---

7 *MJ*-May 12, 1875.

On the evening of Wednesday, May 12, my family sat down to supper without the attendance of company for the first time in days. Afterward, we gathered on the back porch, as was our custom, and I thought it would be a good time to tell them about my discovery at Gowen's house. But once again, Dutch Henry showed up to visit my sister, Mary. He enticed her to walk with him into the back yard, up beyond the stable, and I asked my parents, "How long has this been going on?"

Da' deferred to ma' for the correct answer and she said knowingly, "A few weeks. Mary likes him a lot."

"What d'you think of him?" I asked them.

Da' said, "He's very intelligent, and he's careful, and he can do the work of two men."

"He's a bruiser of a fellow," said ma', "but I sense a kindness in him that I like. He loves music."

"You don't mind that he's not Irish," I observed. "That's unusual in this day and age."

"Can't see that it matters," said da'. "The Irish have some marvelous specimens among 'em, to be sure. But some not so marvelous. Even Christ couldn't have merited salvation if he'd been held responsible for the sins of all of his relatives – a stiff-necked people at best. No, none of us are worth saving except as individuals. It's only ignorant people who revel in their own race. The smart ones know that one race is as bad as another, among those who aren't inclined to try being better than they are."

Ma' chuckled agreement saying, "I'll take a good German over a bad Irishman, any day."

"You should tell the newspapers that," I said. "They're always arguing that the W.B.A. is an Irish union, and for that reason the other races should have nothing to do with us."[8]

---

8 *MJ*-May 19,1875 contains a piece entitled *The Irishman's Union*. *MJ*-May 24,1875 reports that workmen employed at Burnside Colliery near Shamokin were attacked by rioting strikers, and the names of three strikers who were arrested were Mintzer, Adams and Kirby.

"Yes, divide and conquer," said da'. "It's exactly as they did to divide Protestant and Catholic in Ireland."

"I'll be back shortly." I told them as I went up into the yard.

The couple were in an almost level area where da' had suspended a simple board swing under the limb of a huge black walnut tree. Mary was swinging away in happy abandon while Henry watched with delight. He was several years older than I, but we had worked below ground together on several occasions and there was mutual respect.

I whispered to him, "Mary has known some Henrys from time to time, but you're the first who's been German. I call you her Dutch Henry."

He gave me a wry grin and shot back, "Well, I've known quite a few Matthews, but you're the first who's been a Mick. You can be my McMatt."

I chuckled and announced to both of them, "Sorry to interrupt, but I need to talk about something – on the porch."

*   *   *

I related how I'd first met the notorious Jim McKenna in Mahanoy City after the murder of George Major. Then I explained how I'd inadvertently overheard Mr. F. B. Gowen directing his detectives to go after the Molly Maguires, only to discover that his undercover operative within the gang was none other than that same McKenna.

As I talked, I noticed that Henry became increasingly agitated. He grimaced and fidgeted as though he wanted to interrupt me at every turn, but he managed to control himself until I was finished. Then he exclaimed, "My God, Matt, I just saw the fellah you describe! He's right here in Locust Gap!"

Henry explained that he had stopped at Hester's Barroom on his return from his plumbing work at the Ashland House. There, McKenna of Shenandoah had been pointed out to him as a visiting celebrity. He was accompanied by his albino bodyguard and a companion from Girardville. They had taken over the place, buying drinks for everyone, singing and dancing and playing barroom games and being very drunk, although it was early in the evening.

It took me a moment to absorb the information and then I murmured to myself aloud, "What's he doing here? We have no Molly Maguires around here. He must be working with those Coal & Iron Police who just came into Ashland."

"Probably so," said da'. "If I know Mr. Frank B. Gowen, as I certainly do, that detective has been ordered in no uncertain terms to find out who it was who burned down his breakers, and those of his friends."

"Then we had better send this McKenna fellow packing, back to Shenandoah," said Henry.

"Yes, but how?" asked my sister Mary.

We all thought in silence about the answer to that question and Mary added, "He hasn't done anything which would justify harming him."

"I wouldn't stand for that," said ma'.

"I have it," Henry said. "We can threaten him that we found out he's a detective and if he refuses to go back to Shenandoah, we'll tell everyone."

"That could work," said da'.

"Yes," Mary agreed.

"But there's a problem," I objected. "When we tell him we've discovered he's a detective he'll deny it, of course. He'll demand to hear what proof we have. Then I'd have to admit that I overheard his conversation through the vent in Mr. Gowen's basement. It's true that I didn't intend to overhear what I did. But if I now take what I learned and use it against Mr. Gowen, wouldn't I be violating his hospitality, if not his trust? I don't like that."

"It wouldn't be a violation, but I can see there are arguments the other way," said da'.

"Ach, you men and your hairsplitting!" ma' exclaimed derisively. "'What's learned in innocence may be used for any innocent purpose'," according to Caratnia.[9] Still an' all, I'd prefer to avoid losing the friendship of Mr. Gowen if we can help it."

"We don't want to do that," da' agreed.

There was a further silence and then Henry said, "We wouldn't have to reveal that Matt ever overheard anything, if we could prove McKenna to be a detective in a totally unrelated way."

"How could you do that?" Mary asked.

Da' now had a twinkle in his eye as he said, "We give McKenna some juicy information which we know he will have to pass along to that squad of detectives in Ashland, and then we watch to see what happens!"

"That's what I was thinking," said Henry.

"It's a plan that can work!" I approved.

The next day the extended family gathered at Uncle Tom's in Locust Gap, including my oldest cousin Patrick and the rest of the clan from Ashland. We didn't mention my visit to Gowen at Mount Airy, but said only that we had reason to suspect McKenna of passing information to the Coal & Iron Police and we wanted to see if it was true. They were eager to find out as well, for reasons of their own about which no one was so obtuse as to inquire.

"Are McKenna and his men still in the barroom?" I asked.

"Indeed he is," said Uncle Tom. "The three of them slept until past noon, and then went to the bar to eat and to cure their hangovers with the hair of the dog that bit them."

"Good! Good!" said da'.

We divided into teams of two or more men who took turns occupying McKenna in the barroom. We poured liquer into him for four days. The man's ability to consume alcohol was prodigious and he drank several of us under the table. All the while we enticingly supplied him with the information he had come for.

Uncle Tom described for him the details of how the explosive charges had been placed in the timbers of the bridge at Locust Gap

9 Editor's Note: An ancient Brehan lawgiver.

Junction, and explained why the dualin had failed to do much damage, and he swore he could identify the culprits who did it but they had left the area looking for work and he didn't know their names or where they might be contacted. He suffered McKenna to extract this information from him over a long afternoon in which he tried to match the detective drink for drink, and Tom woke up the next day with a raging hangover. McKenna, however, seemed none the worse for wear.

Cousin Patrick, from Ashland, told McKenna a suspenseful story; all about the men from Gordon Plane sneaking past the night watchmen and being lowered down the mine shaft with their axes strapped to their backs, and cutting the cable, and rigging it so it would separate at the precise spot where it would do the most damage to the blacklegs' inclination to stay at work.

"You were there yourself?" asked McKenna.

"No. No, but the man who told me got it from his own cousin."

"Do you know the ones who did it?" McKenna asked.

"Of course."

"Who?"

Cousin Patrick squinted at McKenna as though the answer to his question was too obvious to warrant asking and he said, "Sure, an' they were the same fellahs who were replaced out of work by the blacklegs."

"Who was it that told you the story?" McKenna pressed on.

"Fellow by the name of McGraw. No, wait, it was McGrain. Or was it McGrew? Something like that."

In similar fashion McKenna was told about the burning of the four breakers in which Gowen had such a personal interest. We told him the truth about it – as Pinkerton admits – that the breakers would not have been touched had not the operators continued to put good miners out and blacklegs in, after notification to stop it.[10]

---

10 Author's Note: It's perplexing for an unsuspicious person to read Pinkerton (pp. 286-290) as to how his detective obtained this information. He admits McKenna traveled to Locust Gap to get it but he says it was all forced into the reluctant ear of the detective by a talkative fellow he knew from Girardville, who he took with him the nine miles to Locust Gap. The agency's summary of McKenna's services for

We gave him every detail that an inquisitor would want to know, except anything that might reveal the identity of anyone who could be prosecuted for the arson. We kept giving him details and whiskey with such gusto that I think we might have detained him in the barroom indefinitely. He was having such a good time that on Sunday morning we prevailed upon the owner of the barroom to have his two beautiful daughters take McKenna to late Mass and to dinner and then to keep him away from liquor that day by playing cards with him. We wanted him to be reasonably sober when we hit him with the information that we knew he would have to report, and he was scheduled for a trip back to Shenandoah on the 4:30 p.m. train on Monday.

He came into the bar at noon on Monday to have his dinner, and we were ready for him.

Uncle Tom and I took a whiskey for him to his table and we were invited to sit. Tom then launched into a rambling discourse about how he and McKenna had become great friends over the past week, and there was nothing more sacred than friendship and trust, especially between Irishmen.

"An' it's the last thing in the world I'd want, is to see any harm come to ye, young fellah!" said Tom solicitously.

"Who means to harm me?" asked McKenna looking about suspiciously.

Tom threw up his hands as though surprised he had been misunderstood and said, "Ach, no! You have none but friends around here, Jim. There's no harm meant to you specifically but I'm afraid you might get in the way of what's intended."

"How so, Tom?" asked McKenna.

"Well, me and Matt just came from Ashland. There's some very

(10 continued:) May 9, 1875 reports that he had already gone to Girardsville on that date to ask the same fellow "some questions about the Helfenstein affair". If that man had all that information and was so free to impart it, then why did McKenna continue on the trip to the Gap, one is left to wonder. Pinkerton's book is often wonderful, in that sense. I suspect McKenna never told anyone the truth about how he obtained the information, especially not Pinkerton.

angry railroaders there – fired for joining the union. You'll be all right to take the 4:30 train out, but if you're planning to return at 9 o'clock tonight or perhaps tomorrow, I'd advise agin' it. There's going to be a hell of a wreck, and their target is specifically meant to be the passenger train!"

"Christ!" exclaimed McKenna. "Which is it, Tom? Tonight or tomorrow?"

"Dunno, Jim. They haven't decided, but I know they'd prefer to do it tonight."

"Where? Do you know where they intend to do it?" demanded the detective.

Tom grinned as though with self-satisfaction and he bragged, "Now, that's one thing that I know has been settled for certain. They have all their crowbars and other tools already hidden nearby. They'll rip up the tracks when it's good and dark at 9 o'clock, between Griscome and the New Slope of the West Shenandoah! That's where!"

McKenna looked like a man who's been told he's inherited a fortune. His eyes popped so wide with delight I thought they might roll onto the table. He had to struggle to control himself and keep his voice calm and matter-of-fact.

"Ye're a good friend fer tellin' me this, Tom. You too, Matt. I was planin' to return tonight or tomorrow, an' now I know better."

"But you won't be telling anyone else who doesn't need to know, will ye Jim?" I demanded his assurance. "The way we figure, this is between the workmen and the railroad."

"No. Of course not, Matt," McKenna lied to me convincingly.

He wolfed down his food and got up to leave and I asked him, "Can I buy yez another drink, Jim?"

For the first time in four days he refused alcohol.

"Got to pack now. Later," he promised.

He scurried off to his room. We went into a house across the street that had an attic window overlooking the window to his room. We couldn't see everything in there, but from time to time we could see he was handling sheets of paper. His bodyguard, Whitey, was in there with him. After about half an hour Whitey came out into the street and took off rapidly in the direction of Ashland. One of Uncle Tom's boys followed him and reported that he went directly to the Ashland House, arriving at about 4:30 p.m.[11]

He was inside for only a minute, then he went to the railway station and boarded the train that was carrying McKenna from Locust Gap to Shenandoah.

I also was on that train, and I saw Whitey get on in Ashland where I got off. I went to my cousin Patrick's house and told him the trap had been set. We had only to wait until 9 o'clock to see whether the Coal & Iron Police would be lying in wait along the stretch of track that we had falsely threatened.

"I thought of another thing we could do," Pat told me. "I promised our bellboy friend at the Ashland House that he'd get a dollar for a copy of any telegram which comes in for Captain Linden tonight."

"There might be something," I said, but in truth I admit I dismissed it as a shot in the dark.

Imagine my surprise when at 7:45 p.m. the bellboy showed up to claim his dollar and handed us copy of the following dispatch:

"Shenandoah, May 17, 1875. To: Capt. Linden, Ashland House
IF YOU DO NOT MEET (M) TODAY OR TOMORROW
DISPATCH HERE TO ME. MEET PC ON WEDNESDAY.
LOOK IN THE POST OFFICE. JAMES McKENNA."[12]

"Merciful Mother!" I exclaimed. "The detective used his undercover name to send this! We needn't bother going out to the tracks to prove he's working with Linden!"

But of course we did go out to the area we had threatened, at the

---

11 Author's Note: The report on Captain Linden in Ashland for May 17, 1875 states that he received the letter from "McFarland" warning him of the plot at 4:30 P.M.
12 See summary for Capt. Linden for Mary 17, 1875.

stated time. The summary of Captain Linden's services for Mr. Gowen describes how the Captain and twenty-one men hid themselves deep in the brush along the track, spacing themselves thirty or forty feet apart from each other, eager to pounce upon the expected malefactors when they were in the act. But all they saw were me and cousin Pat and Uncle Tom. The report states:

> "At 9:00 p.m. three men come down the tracks talking; and from their positions the operatives could see them quite plainly as they passed. Operative RJL followed them a short distance; they however passed on quietly and the train passed along all right."

You can imagine the stifled mirth and the back slapping and chortling from the three of us after we had passed out of sight of the big detective. When we were certain we had gone out of his hearing we had to lean on one another for support, we laughed so hard.

The next night and the night after that, the 18th and 19th of May, Captain Linden's so-called flying squad of detective policemen again buried themselves in the bushes in the same location. On those two evenings we sent only one man, Cousin Pat, to observe them, and they duly noted him in their reports on both occasions. You might think that by the end of this time the detectives would have figured out that someone was pulling their legs. But no, they continued in their blissful credulity.

The meeting in PC (Port Clinton) that had been called for in McKenna's telegram of the 17th was rescheduled for May 20. That morning Dutch Henry and I followed Linden onto the train that would take him to the meeting with McKenna. Meanwhile the clan in Ashland brewed up another horrific story to vex the detectives upon their return.

As the dispatch had directed him, the Captain found McKenna at the post office, accompanied, as usual, by his albino bodyguard. The three went to sit openly in the public park. They had probably chosen Port Clinton as the venue for their meeting because it was outside the coal region and there was little likelihood anyone would know them. They had a long and spirited conversation with a lot of hand gestures. The burning of the Excelsior was discussed.[13] Henry and I kept our distance so McKenna wouldn't recognize me. They spent the afternoon in a bar and returned toward Ashland on the evening train. I had seen enough, absent any connection with my trip to Mr. Gowen's house, to confront McKenna with who he was.

But there was more to come. At Mahanoy Plane, a Superintendent of the railroad came aboard the train looking for Captain Linden to tell him the wild story which the lads from Ashland had made him believe. The excited official proclaimed in a loud voice, which Henry in the same car could easily overhear, that hordes of striking miners were assembling from Mahanoy Plane, Girardville, Ashland and other places with the determination of marching over the mountains and preventing all the collieries in the Shamokin region from working.[14] Linden was quite alarmed and sprang into action. Then and there, on the train, he dictated telegrams for the railroad official to have transmitted as soon as possible. The dispatches ordered the flying squads to assemble and directed them to positions from which they could best protect the collieries and keep a sharp lookout for the marchers coming through the mountain passes. Of course, Henry and I knew there were no marchers. When we thought of the thirty or more detectives who would be lurking in the woods all night looking for mobs who were mere phantoms of the lads' imaginations, we laughed so hard that our sides hurt.

It was one of the precious few laughs to be afforded us in that nasty business. I didn't foresee how deadly it would soon become, probably because I had not yet brought myself to accept the fact that evil engorges the hearts of some men.

13 Author's Note: Pinkerton (pp. 293-297) recites that on May 18 and May 19, McKenna learned that the men of his division were planning to murder the Welsh pistolero Gomer James on May 20, but he couldn't warn anyone because he was watched too closely by his men. Yet we know McKenna must have communicated with Linden on the 18th or 19th to reschedule their "PC" meeting from the 19th to the 20th. Also, Linden's report of the Port Clinton discussion makes no mention of being warned about the plot to murder James. Clearly, Linden was not warned of it, at least for the purposes of the Agency's records, and McKenna was making no effort to save the Welshman from his men. I personally suspect that McKenna rescheduled the meeting for the 20th so he would have an alibi should the ambush go as he had planned. In the event, the Welshman didn't show up.
14 Linden summary for May 20, 1875.

Chapter 9

McKenna Sent Packing

McKenna and his shadow, Whitey, got off the train in Girardville, presumably to investigate our false report about marchers assembling there. Dutch Henry and I followed. They went into the telegraph office. I sat myself on a bench opposite the door where they would be certain to see me when they came out. Henry, whom neither of them had met, lounged across the street where he could observe discreetly.

"McWilliams! What are ye doing here!?" McKenna seemed angry to be found coming out of the telegraph office.

"Looking for you." I told him, remaining seated and offering no handshake.

"What's going on?" he demanded and looked around to see if anyone was with me.

"We need to talk," I told him, and added with a meaningful glance at Whitey, "in private."

He motioned his bodyguard to move away and sat down on the bench as far from me as he could, and stared at me, saying nothing, waiting for me to begin.

"The lads at Locust Gap have found out that ye're a detective," I announced bluntly.

His jaw dropped in surprise and his first reaction was denial. "Who told ye such a lie!? Who!?"

"Jim, they know!" I simply insisted.

"If ye say that again I'll beat the crap out o' yez!" he threatened.

I looked him in the eye and explained, "You're the only one who was told about the plan to derail the passenger train, and the place was swarming with police."

"Well, the police must have found out in some other way, but not from me," he argued.

"Jim," I patiently explained, "there was no actual attempt. Ye were told a story to see if ye'd pass it along, and ye did."

It took him a moment or so to process that challenge, but then he strove to overcome it.

"That doesn't mean I'm a detective," he argued. "I just couldn't stand by and let all those passengers be injured, is all!"

"Ye peddle a slick line of goods, Jim," I complimented him. "But we saw ye spend all day in Port Clinton with Captain Linden. There's no denying it."

"Ye followed us?! What did ye hear us talkin' about?" he demanded to know, and for the first time there was fear in his voice.

"I'm no eavesdropper!" I defended myself because I still felt guilty about listening through Gowen's ash vent. "I don't try to overhear people's private conversations. If I do, it's by accident. I don't know what ye talked about."

He seemed very relieved to hear this, and immediately I realized I had provided him with too much information, which he used to advantage.

"Well sure, the Captain had me meet him in Port Clinton," he admitted. "An' he spent the whole day tryin' to convince me to become an informer for him. But I turned him down flat, boyo, and there's no cause for sayin' that I'm a detective."

I was thoroughly fed up with his lies, but I held my temper and said, "But if ye had never met him before, how did the Captain manage to set up the meeting? We know your man Whitey never spoke with him, but simply dropped off your letter."

"My letter told him where to telegraph me in Shenandoah an' that's what he did."

"Oh, I see!" I said sarcastically, "Ye sent the police Captain on a wild goose chase to foil a plot that didn't exist! An' he telegraphs you, a complete stranger, to meet him in Port Clinton so ye can sell him more bogus information! Is that about it?!"

McKenna shrugged off my sarcasm and said brazenly, "Ye can't say that such things don't happen."

"All right, Jim, let's peel away the last lie," I said to him pointedly and asked, "What do the initials 'P.C.' stand for?"

"Port Clinton!" he answered defiantly. "We jist came from there."

"Why wouldn't 'P.C.' signify Port Carbon?" I challenged him. "That place is closer to here, and is still in the coal regions."

"Well, 'P.C.' might also signify Port Carbon," he admitted.

"But it doesn't," I announced, "You and Linden had arranged in advance a code in which 'P.C.' means Port Clinton, so that when ye sent him that telegraph to meet in 'P.C', he knew exactly where ye meant! Do ye want to see a copy of your telegraph, McKenna? The original is at the Ashland House!"

"Oh, Christ!" he exclaimed, for he knew his identity as a detective could be proven. Then he snarled at me, "Ye lousy son-of-a-bitch!" with such intense hatred that I sensed he was capable of murder should it suit his purpose.

"It's not only me," I reminded him of why he shouldn't act rashly. "Several of the Locust Gap lads know about it, Jim."

He nodded to let me know he understood and said, "They sent ye to talk to me. Why?"

"Because they might want to let ye keep your secret. It depends," I said.

"On what?"

I made a mental note that the only time he seemed genuine was when he was asking questions.

"It depends upon yer answers," I told him and asked, "Who's yer employer?"

"The Pinkerton Detective Agency."

"Who hired them?"

"The Railroad."

"Why are ye here?"

"To root out crime."

"Crime against the Railroad?"

"No, crime against anyone, mostly by the organized gangs, the Modocs and the Molly Maguires."

"Then why would ye come to the Shamokin area? There's no Molly Maguires there," I argued.

"I know that now," he agreed. "But with all those breakers being burned and collieries attacked, I was sent to check it out."

"What's happened around Shamokin is because people were tryin' to save their jobs, not because they are criminals," I argued.

"Fair enough. But that isn't true around Shenandoah and Mahanoy City," he said. "Have ye heard the latest about Dan Dougherty?"

"No," I admitted.

"He wasn't back in town out of jail for as much as a full week when he was shot at to avenge the Burgess; they say by Jesse Major an' another Modoc by the name of Bully Bill Thomas. He's as rough a customer as ye can get."

"That's preposterous!" I exclaimed. "Dougherty was conclusively proved to be innocent of the shooting of George Major!"

"I know. I read the newspaper report of his trial," said McKenna. "But innocent or not, the Modocs want him dead."

"That's dreadful," I said.

"It is, but we have much the same on the Irish side," said McKenna. "You've heard about the shootist Gomer James who was acquitted of the murder of the Cosgrove boy. Well, the Cosgrove family has put a price on Gomer's head, and the Mollies are out to collect it."

"Oh, I see."

"I'm not tryin' to imprison angels," he added indignantly. "An' it's a damned dangerous thing."

"Exactly who are these Mollies that people talk about?" I asked.

"We've never met anyone, not a single person ever, who's called himself a Molly!"

"No an' ye won't," he said. "They're an ultra-secret clique, an' they use many names."

"That's stupid!" I exclaimed. "If they don't call themselves Mollies, then you have nothing more than a name made up by someone else. A fiction, useful only to confuse people."

McKenna looked at me as though surprised that I had understood some deep enigma, but he quickly recovered his composure and said, "I won't argue about the name being fiction, but I know the gang is real."

"Then tell me, what is its connection with the Ancient Order of Hibernians?" I demanded. "The Bishop and some of the priests talk as though the A.O.H. and the Mollies are one and the same."

McKenna grinned slyly at me and said, "Not all A.O.H. are Mollies, but most of the Mollies are A.O.H."

"Sounds like more fiction," I said bluntly. "I know that John Kehoe was installed as the A.O.H. delegate for Schuylkill County in order to root out the lawlessness which had been permitted by his predecessor who is your particular friend, Barney Dolan. An' he's issued strong words threatening to expel those who would use the A.O.H. for evil. Isn't that so?"

"It is," McKenna acknowledged. "But now the word is out that the operators in Schuylkill are going to open their mines using blacklegs like the Shamokin operators have been doing. If that happens there's going to be hell to pay, and neither Kehoe nor anyone else will be able to stop it."

"Even so, that won't have anything to do with an organization called the Molly Maguires." I got back to the point.

"Yes it will," argued McKenna. "The men of Schuylkill will use the A.O.H. organization to resist the opening of the mines an' a lot of the A.O.H. Bodymasters are Mollies."

"So you're saying that in some places the Bodymasters are in control and some of them are Mollies."

"Correct," McKenna approved and added the phrase from the Bible, "An' by their fruits ye shall know them."

"Then the members of their divisions must know who they are," I observed.

"Oh, no! Oh, no!" objected McKenna. "They never discuss their special Molly Maguire business in open meetings. If a job's to be done, the Bodymaster speaks of it only to the men who he picks to carry it out. Never to the others. The members of the lodge are never made aware of the job or the identities of those who do it."[1]

"That means there's no actual organization," I said. "Rather, there's an accidentally related cabal of people who occasionally conspire to commit crimes."

"Almost all of them are Hibernians," he protested.

It wasn't my purpose to convince him of anything, so I dismissed his statement with a depreciating wave of the hand and said, "We don't have any objection to ye fighting crime. But throwing people into jail for demanding decent wages and working conditions is another matter entirely."

McKenna instantly understood and he declared, "The Pinkerton Agency chases real criminals, like Frank and Jesse James, the Younger Brothers, and the Reno gang. You read the papers. When your Mr. Siney and Mr. Parks were arrested for resisting those Italian strikebreakers up in Clearfield County, we had no part in it, did we?"

"Not that I know of," I admitted.

"Well, we didn't, or y'ed have read about it in the papers," he insisted. "I'm a working man myself, though not a miner, an' I'm tellin' ye I wouldn't have any part in it. An' if the A.O.H. goes out marchin' to keep the West-Shenandoah colliery closed to strikebreakers, I'll be out there marchin' wid 'em!"

Based upon that promise, and his pledge to never again set foot in the Shamokin region, McKenna's identity as a Pinkerton detective was not revealed. The family agreed that it might be best to keep

1 See: Report on operative McKenna for March 24, 1874.

intact the devil we knew rather than have him replaced by one we didn't.

But I warned him in no uncertain terms that I, and some others both in and outside of the W.B.A., would be watching him and judging whatever he did. Only now, with hindsight, do I fear that the warning may have had the unintended consequence of engendering in him a more refined perfidy.

Chapter 10

Assault On Bully Bill

When I returned to W.B.A. headquarters I learned that McKenna had been right about the next big crisis in Schuylkill County being caused by the operators trying to keep open their mines with imported blacklegs. Such use of scabs had effectively demoralized the strikes of the loosely federated, single mine unions in the Shamokin area.

This was the W.B.A.'s opportunity to demonstrate the superiority of our unique form of organization, in which the men of each colliery deferred to the central union to guide it, and in return received support from the entire county, and even other counties.

Our plan was very bold. We would hold picnics, dances and public orations to mobilize the miners and their families. Our men would march the public roads, going from colliery to colliery and peacefully but effectively urging all, especially those who were working, to join the strike. The men of Schuylkill would even be joined by marchers from Luzerne County, in and around the city of Hazleton. The strength and solidarity of the W.B.A. would be proven in a very public way, and the mines would be closed, and the strike continued, all without violence.

Our most important advantage, and well did we know it, was that the men had a common grievance against a common enemy, and it united them. As Mr. Pinkerton's book at page 329 put it:

"Among the miners, the Welsh, English, German and Poles

mingled, and heartily joined hands with the Irish. For once, feuds were forgotten, and nationalities all made common cause."[1]

My job was to go to Girardville and ask John Kehoe for his help. He had recently been elected High Constable of Girardville. He was the most prominent man in the Girardville-Lost Creek-Shenandoah area, and he had already demonstrated in Mahanoy City that he and his A.O.H. Bodymasters could stage a perfectly peaceable march.

"But don't deceive yourselves that this won't be a different kind of march, Matt," he warned me. "Ye'll have opposition from those already at work, and their families, and from the Coal & Iron Police, and from people who are always against any kind of confrontation because they feel threatened by it."

"We know," I told him.

"And the newspapers and the authorities will hold ye responsible for any violence no matter who commits it."

"We know, but they can't arrest all of us," I argued.

"Alright," he agreed. "My Bodymasters will help with the organizing an' all, but it must be a union march. We don't want it to seem that we're in charge."

"Thanks, Mr. Kehoe," I said gratefully. "Our President also sends his thanks."

"Save yer thanks, lad," Kehoe smiled craftily. "There's something I want in return. From yerself, personally."

"From me?! What?"

"Yer friend Dan Dougherty needs help. There's some in Mahanoy City who still believe he shot the Chief Burgess. He can't walk down the street without getting bullet holes in his coat."

"Who?" . . . I began to ask.

"Who knows?" he interrupted. "There's at least fifty men in Mahanoy who might want to shoot him. Almost that many testified

---

1 Author's Note: This contradicts the claim at page 302 of Pinkerton's Book that Kehoe found the English, Welsh and Irish in Mahanoy City at each other's throats, and that gave him his motive to plan murder.

against him. But the one with the biggest mouth is a pig stealin'
Welsh brawler called Bully Bill Thomas. Dan's especially frightened
of Bully Bill, 'cause he's crazy. But Dan is also fightin' mad. He won't
run. He wants protection an' he's going to make a formal request to
me, as the A.O.H. County Delegate, to help him. I don't see how I can
say 'no'."

Kehoe stopped to let his words sink in, and my first thought was
that the union's hard won worker's solidarity would be shattered by
a clash between the A.O.H. and Bully Bill and his ilk.

"This couldn't happen at a worse time," I said, "D'you think we
could convince Dan to leave town for a while?"

"No. He's too proud an' too stubborn. He's already been asked,"
Kehoe replied.

"Then, what the blazes can I do about this mess?" I asked in
frustration.

"When Dougherty makes his formal request for my help, I want to
be able to promise him some action," answered Kehoe. "Ye're studying
law, and ye were on the defense staff for the whole of Dan's trial. I'm
thinking ye could make a summary of the evidence that proves that
Dan is innocent. We can have it published in the *Miners' Journal*. That
might make him safer on the street, or feel safer, at least."

"That's a good idea," I agreed. "The trial transcript was too much
reading for most people. A summary is good. But it would carry more
weight if it came from the defense attorneys themselves. I'll draft up
something and see if I can't get them to sign it."

"It'll be a big help to Dan and to the strike effort, both, if ye can
do that," said Kehoe hopefully.

I did do it. My draft was reworked by the lawyers and botched by
the paper's printer to be less compelling than I had hoped, but it is
nevertheless a remarkable document. Here is an excerpt from the
Miners' Journal:

THE MAJOR ASSASSINATION –

A DECLARATION OF BELIEF IN DOUGHERTY'S INNOCENCE

Editor *Miners' Journal*; The late attempt at the assassination of Daniel

Dougherty, it is perhaps just to believe, could have been instigated only by the supposition that in his recent prompt acquittal of the charge of murdering George Major, an offender had escaped the penalty of the law . . . .

The undersigned, therefore, the counsel of Daniel Dougherty . . . hereby assert our fullest conviction of his entire innocence as to the death of George Major. The witnesses who testified to Dougherty's identity as the person who fired the fatal shot . . . were certainly mistaken . . . The evidence of the surgeons alone fully established that Dougherty could not have been in the position of the man that George Major shot, and who it was admitted on both sides was the one who shot George Major . . . Dougherty was also shot so badly that he expected to die from his wound. He received . . . the last rites of his religion. . . Then in . . . preparation to meet . . . his God, Dougherty declared his innocence. To this best test of the real truth of the case, we add . . . our personal declarations, that we are fully convinced that Daniel Dougherty is entirely innocent of the charge of having shot George Major, and we affirm the same conviction that he in no way aided, encouraged or assisted in such shooting.

F. W. HUGHES
HORACE M. DARLING
O. P. BECHTEL

Getting three busy lawyers to agree upon the precise wording of the notice delayed its publication until June 7, 1875. That was only six days after Dan Dougherty formally applied to the A.O.H. for help. So it was a fairly prompt response. After it was published there were no further attacks on Dan's life.

*   *   *

In telling how I helped to keep tempers cooled in spite of the shootings at Dan Dougherty, I've gotten six days ahead of my story about the county-wide, united workers' demonstration against the use of blacklegs to keep the collieries working.

Kehoe and his A.O.H. leaders from all over the county and beyond, helped the union to successfully stage the workers' march as planned.[2] There was a wonderful picnic and dance and daytime parades by more than two thousand men, and it was almost entirely peaceful. The militia was brought into Mahanoy City but they proved unnecessary – the reaction of policitians to a press hysterically biased against any kind of union activity.[3] Of course, my own bias in favor of it must be duly noted.

In any event, no one denies that the day before the demonstration was to begin, June 1, 1875, Kehoe and his A.O.H. leaders convened at Clark's Emerald House in Mahanoy City. His men came from as far east as Coaldale near the Carbon County line, as far west as Locust Gap in Northumberland County, as far south as Mt. Laffee near St. Clair, and as far north as Shenandoah, where the division leader was Jim McKenna. It is fairly clear that the assembly of so much talent could be justified only by a very compelling purpose, such as coordinating a workingman's parade. Similarly, it's clear that Dan Dougherty was permitted to appear before the assembly, and he held up his bullet-riddled coat and asked for help.

It is not disputed that Jim McKenna attended the meeting as the Shenandoah A.O.H. leader; that he ordered a young man to take down the names of all who attended; and three days later, after the workingman's demonstration was over, he went back to Shenandoah and ordered the men of his division to go out and kill Bully Bill Thomas. It took them awhile, but they made the attack on June 28. It was McKenna's first documented crime.[4]

What has been hotly disputed about that June 1, 1875 meeting of the A.O.H. is McKenna's claim, made a year later, that Kehoe's only

---

2 Pinkerton, pp. 328-332
3 MJ-June 4, 1875
4 Comm. v. Kehoe, re: Wm. Thomas, pg. 63;McKenna said that two brothers of George Major were also targeted, but no assault on them was ever made.

reason for calling the meeting was to give the order to McKenna and some others to go out and kill Bully Bill. And that was the only business transacted at the meeting, according to McKenna.[5]

Had I told Kehoe at the time that McKenna might make such a claim against him, he'd have laughed and said, "Let him make it! It's so stupid no one will believe it!"

On the face of it the claim was outrageous. Why would Kehoe call a county-wide conclave to hatch a local murder plot which he could have whispered secretly into McKenna's ear only? And the alleged plotmaster Kehoe, according to everyone including McKenna[6], had been trying for months to rid the A.O.H. of crime. And the timing is profound. For the first time in memory the Welsh and the Irish and the English and the Dutch were all united. It is then that the masterful politician Kehoe, whose power depends on success of the strike, allegedly sent his most notorious gang of Micks to go out and murder the Modocs! In the middle of a planned solidarity demonstration! I lose my breath when I think of the mind-boggling audacity of the charge. Only those who desired to fracture the union and paint it as criminal would have had a motive for such a murder.

But oh, you who might be tempted to cast your fortune upon public opinion and the roll of legal happenstance, take heed. Take heed of what history already knows. Every one of those leaders who attended Clark's saloon that day except McKenna, and one of his henchmen named Kerrigan, and another who fled the indictment, were found guilty as charged. Guilty!

So it was. I have no provenance to criticize the defense lawyers for not explaining to the jury what was the real purpose for the June 1 meeting. It's understandable, they didn't want to admit that the defendants had participated in acts which had caused the calling out of the militia. For the same reason, they wouldn't let me testify that prior to June 1 Kehoe had sent me out to get Dan's lawyers to solve

5 Comm. v. Kehoe, re:Wm. Thomas, pg. 29.
6 Report on McKenna for Ap. 29,1874 and May 6,1874.

his problem without violence. They were afraid that on my cross-examination the A.O.H.'s connection with the union would come out.

I must now ask the reader's indulgence to skip forward in my narrative to the trial of Kehoe and his A.O.H. leaders for ordering the murder of Bully Bill. We must discuss detective McKenna's testimony which contains serious admissions, as well as some conflicting evidence which proves the truth of what occurred at the meeting and what happened to Bully Bill.

It was the most famous of all the Molly Trials. The prosecuting attorney was none other than Railroad President Gowen, himself. He formally promised to prove that John Kehoe called the meeting of A.O.H. leaders for the purpose of ordering McKenna and others to murder Bully Bill and two brothers suspected of shooting at Dougherty. And further, he formally promised to prove that the resulting attack by the detective's division on Bully Bill was made in pursuance of the rules, regulations and orders of the A.O.H.[7]

By this time, I was working as a law clerk for one of the defense counsel, and had learned enough to follow the testimony with a critical ear.

McKenna was sworn in. He testified that he came into the coal region and joined a secret organization named, "The Ancient Order of Hibernians, more commonly called Molly Maguires." (p. 16). He then proceeded to describe how the A.O.H. was organized from its local divisions to the National Board, and gave a long dissertation on the signs and passwords by which the Order's members could identify one another. (pp. 16-22).

Then Gowen asked:

"What was the practice of this organization in reference to committing crime?"

Answer: "It was a general practice to commit crimes." (p.22).

---

7 Comm. v. Kehoe, re: Wm. Thomas, pg. 16. The text references to pages are all references to the transcript in that case.

At this point, had I been counsel for the A.O.H., I would have stood up and yelled, "Objection! Such testimony must be stricken! The best evidence of the rules and practices of the A.O.H. are its written By-Laws and Constitution, present here in Court. Those documents clearly prohibit the commission of crimes!"

Of course, the A.O.H. was not represented by counsel, so no such objection was made.

Later, on cross-examination, McKenna was forced to admit that the criminal practices he had described were contrary to the A.O.H. Constitution (pg. 55). Therefore, Mr. Gowen failed utterly in his promise to prove that the attack on Thomas had been pursuant to A.O.H. rules and regulations.

Mr. Gowen proceeded next to ask McKenna how the crimes of the A.O.H. were generally committed and the detective answered,

" . . . the division master of that district (where the crime was to be committed) would either apply to another division or to the county delegate to get men who were unknown to the parties on whom the outrage was to be perpetrated."

Gowan's face registered annoyance. This answer suggested that crimes were secretly arranged between division masters, or between a division master and the county delegate. Only a few men involved. In the case being tried, Attorney Gowen was claiming that an assault had been openly planned at a county-wide meeting. A secret arrangement between two division masters wouldn't serve his purpose.

Gowen cleared his throat loudly and asked the question again, this time suggesting the answer.

Q. "In what manner would these divisions determine on the commission of crime? Would it be in meetings of the organization?"

Clearly, Gowen wanted the detective to answer that crimes were arranged in open meetings, and not in secret by certain cliques. And still, the detective's answer was less than satisfactory.

A. "They would have a meeting; sometimes all the members would not be present."

Gowen's face grew red with frustration. The answer can be interpreted as meaning that at the times when crimes were planned, all the members would not be present. To get the answer he wanted he asked another improper leading question:

Q. *"Sometimes all the members would be present, and at other times they would not all be present?"*

A. *"Yes, sir."*

In this manner, attorney Gowen succeeded in getting the witness to suggest that crimes were generally planned in open meetings, although sometimes not all members would be present.

In hindsight, this exchange is very revealing of the character and credibility of both McKenna and Attorney Gowen. After the trial had ended and could not be reopened, I discovered that when McKenna was still out in the field, allegedly among the Mollies, his reports on this very subject were as follows:

Report for March 24, 1874:

" . . . whenever a 'job' is to be done, all A.O.H. members are not notified of it, as some might squeal. But that a certain party or committee was selected and appointed and notified what to do and how to do it, in order to insure secrecy and safety to themselves."

Report for April 30, 1874:

"The operative states that . . . when there is a 'job' to be done - man to be beaten or murdered – the question or matter is NEVER brought up in open lodge, but the Bodymaster . . . appoints the man or men privately and secretly notifies them of what they are required to do. And thus the 'job' is done and the very members of the lodge are never made aware of the transaction or who the 'avengers' are – which must be kept a profound secret by the principals . . ." (Emphasis added).

The detective's reports to Gowen flagrantly contradict the above trial testimony which Gowen had to drag out of McKenna – that decisions to commit crimes were made in open meetings. The conclusion is inescapable that Gowen was an unethical lawyer and that McParlan, when reminded of what was needed, would readily lie under oath!

As to whether the Mollies and the A.O.H. were one and the same organization, the written field reports destroy the testimony of the only witness who ever said they were. Furthermore, the written reports and not the trial testimony has the "ring" of truth about it. People who want to hire murderers and stay out of jail hardly ever call county-wide meetings to appoint the assassins.

I sometimes lie awake at nights, filled with guilt for not having discovered and copied those sections of the detective's reports until after the trial was over. I believe that had the jury been shown these contradictions of McKenna's testimony – though they had been shorn of Irishmen and predisposed to find guilt – they would not have believed that John Kehoe ordered the death of Bully Bill in an open meeting. Unfortunately, we could not convince the judge that this evidence warranted his setting aside the verdict and ordering a new trial.

*   *   *

The day after the June 1 meeting, at which McKenna claims he was ordered to kill Bully Bill, he met with Captain Linden. The detective was supposed to keep the Captain abreast of everything that was going on. Linden's report to the Agency for that day does not even mention the June 1 meeting, let alone a murder plot.[8]

To be sure, there was a plot to murder Bully Bill, but it belonged solely to McKenna and Captain Linden and their henchmen.

Linden had moved his headquarters to Shenandoah and the Coal & Iron policeman had become the "hail fellow well met" companion and drinking buddy not only of McKenna but also of his band of Mollies. Linden often bought drinks for the whole gang and, as Pinkerton put it in his book, " . . . the Mollies did not wonder, or indulge suspicion, when they saw Linden and McKenna occasionally in company." In fact, Linden had convinced the gang that he

8 Report of Linden for June 2, 1875.
9 Pinkerton, pp. 341-344

wouldn't arrest them no matter what they did.[9] Understanding this, it is difficult to believe McKenna's testimony at the trial that Linden wasn't aware of the gang's three-week odyssey to murder Bill, and that the Captain couldn't be warned to prevent the deed because McKenna couldn't be seen talking to him.

The target of the plot, William "Bully Bill" Thomas, was one of the more unsavory Welshmen of the region. He had been part of a pig stealing ring down in St. Clair and had fled to escape prosecution, only to reappear in Shenandoah as a fairly successful bare-knuckle boxer.[10]

Shooting and brawling and threatening murder added to his celebrity, which increased the size of his purses. Perhaps for this reason, along with others, he had taken to loudly declaring that Dougherty should be shot.

McKenna's admissions at trial about hunting down Bully Bill would have landed most mortals into the pokey.[11] On June 5 he personally led a squad of killers to seek for the prey, but he called off the attack because the militia brought in to stop the workingman's march were still thick on the ground.

On June 10 his roommate, Mickey Doyle, and his most practiced cutthroat, Tom Hurley, went together to search for Bully Bill. When they hadn't returned by June 12, McKenna went looking for them. He found them in a boarding house. They reported no luck in finding Bully, but said they were out of work anyway and had the time to spare. The two of them returned to Shenandoah on June 15 and June 16 to report to McKenna that they had lain in wait to ambush Thomas, but the Welshman never appeared.

At this point McKenna claimed to have become ill. On June 23 while he was sitting up in bed, an informant came to him to report where Bully could be found, working the day shift at a colliery.

The next day the technical Bodymaster of McKenna's division,

10 Editor's Note: See: *St. Clair*, pg. 535
11 See Comm. v. Kehoe, re:Wm. Thomas, pp. 31-38

McAndrew, returned to town, and McKenna claims that he himself was no longer in charge of the operation to murder Bully. But the gang continued to think he was because they all continued to report to him.

They met on the porch of his boarding house the afternoon of Sunday, June 27, and with him they made plans to finish off Bully Bill the next day. McKenna gave to Tom Hurley his own coat, a long, whitish-gray slicker, to wear as leader of the squad.

McKenna had all of that evening to get Captain Linden to prevent the impending murder or to warn Bully Bill, but he didn't do either.

After half-past seven the next morning, the death squad returned to report to McKenna.

He met with them in the woods. They told him that Bully Bill was dead. McKenna went back into town and ordered some men, including his technical boss, McAndrew, to bring bread, butter, ham, pie and cigars to a victory feast in the woods. McKenna brought the whiskey.[12]

But McKenna's four assassins had been cruelly deceived, as the detective soon learned.

A little after five o'clock on that morning of Monday, June 28, 1875, a group of seven men got off the Lehigh Valley freight train near the colliery at Shoemaker's Patch, a mile and a quarter north of Mahanoy City. The *Miners' Journal* snidely stated that they were "Greeks or Mollies," so secure was it in its belief that people's nationalities were revealed by their faces. Four of the men walked over to the mouth of the mine and casually sat down. It was the custom for laborers, even strangers, to hang out at the drift-mouth looking for work. No one would challenge them if they didn't get in the way.

From where they sat, the four had a view of the patch itself – the little cluster of houses for the colliery's work force, where Bully Bill Thomas lived. They knew which house was Bully Bill's and they

---

12 Comm. v. Kehoe re: Wm. Thomas, pp.31-38

watched intensely for the Welshman to appear.

Their leader, Tom Hurley, was pug-ugly and sinister, somewhere in his twenties but seeming older, his habits of dissipation and violence showing in his visage and his attitude. He had been an intimate of McKenna from the very beginning, a self-confessed assassin, sometimes residing in McKenna's boarding house and stashing his weapons with his landlord.[13] The only act of human kindness ever recorded of Hurley was his gift of a blackjack to McKenna to use upon the toughs in Scranton.[14] No one in Shenandoah would have believed that he was not McKenna's man.

He sat up straight and alert as he saw Bill coming out of his house in work clothes, carrying no weapons, only a dinner pail.[15]

Hurley shoved his elbow into the ribs of the thin young man next to him and whispered, "There he comes!"

"Ow! I see him, fer Christ's sake!" protested Mickey Doyle.

Doyle was a cringing little weasel of a fellow, not much over five feet in height, always eager to make himself useful to the tough guys he linked up with, and always pleased when that involved picking on someone else. He was somewhat of a vagrant and he was very proud that McKenna allowed him to bunk in his room. The detective could not have had many secrets from his adoring lapdog Mickey.

They watched as their quarry made his way to the colliery stable that sheltered the horses and mules. Thomas was a hostler whose duty was to care for the animals. The double doors of the stable were wide open, as was the adjacent man-door, and Bully went in.

"We'll take him inside while he's at work," said Hurley getting up. The others followed. They walked toward the stable slowly, to avoid arousing suspicion, but their appearance alone was sufficient to warrant notice.

Hurley wore his gang leader's whitish-gray full-length coat and a low-crowned whitish-gray slouch hat, reminiscent of the fabled garb

---

13 Pinkerton, pg.304
14 Pinkerton,pg.189
15 The account that follows may be confirmed in Pinkerton, pp. 325-326 and in Comm. v. Kehoe re: Wm. Thomas, testimony of Thomas and Doctor Bissel.

of the Whiteboys in the old country. It wasn't Hurley's style to do such a job in daylight wearing a white coat, but that's what McKenna wanted. Inside the coat pocket, in his hand, he held a large, silver-plated pistol. Though it was a warm morning, the others also wore coats, dark gray or dark blue, not unusual for work in a mine. What was unusual about them was that they weren't carrying dinner pails, and all kept at least one hand in a coat pocket, and their stride, although slow, was purposeful.

Bully Bill had noticed them sitting at the drift-mouth when he'd gone into the stable to work, but he hadn't been warned and thought little of it. He became occupied with receiving the day's instruction from the stable boss.

As Hurley walked to the stable he worried whether he could really count on Gibbons and young Morris to do their part. Both their families were the "respectable" sort and looked down with snooty disapproval on their association with himself and McKenna. Both boys had been turned sour on mining by the failure of the strike, and they blamed their lackluster lives and prospects upon bad luck, anti-Irish bigotry, unfair bosses and everything but their own fault and their fondness for hanging out in barrooms. McKenna had become their mentor, with his big talk and easy money. They were full of hurt pride and bravado, mindlessly convinced that the murder of the obnoxious Bully was justified by his own bluster about violence.

Hurley would try to get them to do most of the shooting – sacrificial goats should anything go wrong – though he doubted they had the stomach for it. No matter, for he was convinced that McKenna's special understanding with the big bull, Linden, would cloak him with whatever immunity might be required. It would be sufficient for his purposes if he made sure that the two rich kids merely fired their pistols.

They arrived at the stable and Hurley told them, "You two go in through the wagon entry. I'll take the man-door. Mickey'll stand guard. Blaze away the moment ye see him!"

Inside the stable, Bully had two horses and a mule he was leading into the blacksmith shop. He stood with his right hand on the mane of one of the horses, talking to the stable boss with his back to the open outside doors. The boss' face turned ashen and in mid-sentence he stopped talking and bolted away into the blacksmith shop.

Bully swiveled his head toward the doors to see what had caused the flight and there was a man in a white coat and hat pointing a big silver pistol at him. It fired. Bully felt a searing pain as the bullet hit his ribcage on the side and passed downward along the skin and out his lower abdomen. His fighter's instinct to close with an opponent took over. In one swift motion he let go the horse's neck, grabbed his hat from his head, and sent it sailing at the shooter's head, causing him to miss with the second shot. In the same motion he lunged forward and grabbed the silver pistol and held on.

He and the man in the white coat whirled around and about in their struggle for the weapon and Bully saw other men with guns outside the double doors. His opponent was very powerful. The fellow managed to cock the pistol and fire it only inches from Bully's face, the bullet creasing his hand on the barrel and the blast burning his cheek and blinding his eye.

He let go of the pistol and staggered back with his hand over his burned eye. Shots were fired. Horses and mules screamed with fright and with pain. He was wet with blood. With his good eye he saw the silhouettes of men shooting at him and felt the searing of a bullet across his neck, upwards. Dazed, he focused upon a shooter holding a black pistol, shakily trying to aim it, his face as pale and frightened as though ready for the coffin; more scared than Bully himself.

The fellow closed his eyes and fired and Bully felt himself hit again, this time lower in the neck below the clavicle. He collapsed to his knees and crawled away under the terrified animals, heedless of their stomping and flailing hooves. At that moment the horse whose neck he had been holding, which had been struck by the first bullet fired along his ribs, collapsed on top of him. Bully was pinned by the carcass to the floor, dazed, breathless, half blind and helpless. He was soaked with blood.

He lay still, but out of his one good eye he could see a man, a smallish one, framed in the doorway, shakily pointing a pistol at him.

The man in the white coat screamed, "Shoot him ye goddamned coward! Shoot him!"

Bully held his breath. The man shot and the bullet hit the mule in the leg. The animal screamed and kicked wildly. The men left.

Bully forced himself to lay still, but only for a minute. He knew the assassins would be in a hurry to leave. It took all of his strength but he pulled himself from under the horse. He got to his feet and staggered over to the stable door in time to see the death squad going up the hill in the direction of the mountain path to Shenandoah. He was madder than hell.

* * *

As the *Miners' Journal* of the next day reported, a doctor from Mahanoy City came at once to dress his hurts and pronounced them not to be life threatening. Thomas, it alleged, ". . . has incurred the enmity of the "Greeks" in Mahanoy City and this was doubtless the execution of a determined plan to kill him."[16] At the time, I thought this allegation of the newspaper to be questionable because this botched attempt appeared nothing like the "execution of a determined plan" but rather a spontaneous or at least poorly devised assault by amateurs. If the writer of the article actually knew of McKenna's "determined plan" to kill Bully Bill, then he knew as much and more than had been put in the reports of the Pinkerton Agency. As Linden himself had said, there was much conjecture that the *Miners' Journal* was being run by the Railroad.[17]

Anyway, when I read the article, I assumed no connection between the June 1 assembly of the A.O.H. and the attempted murder of Bully a month later. Since the issuance of our statement about Dougherty's innocence he had been left alone, and Bully was always in and out of one scrape or another.

---

16 *MJ*-June 29,1875
17 Report on Linden for May 22, 1875.

The *Miners' Journal* continued to cover his escapades. In its June 30, 1875, issue it gloated upon his manly courage with the following blurb:

"On Monday the citizens of Mahanoy presented William Thomas, alias Bully Bill, with a Smith & Wesson revolver, and yesterday morning he went to work with it in one hand and a dinner bottle in the other."

A month and ten days later the *Journal* conceded that their courageous Bully had taken his lethal new present with him to Port Carbon and opened fire on a barber's pole. Four days after that it had to report that he and another man had engaged in a gunfight on the streets of Mahanoy City and an innocent bystander named Zimmerman had been killed by a stray bullet. The death was deemed by the authorities to be by accident and no arrest was made for it.

I knew only what I read in the papers, and that I took with a grain of salt. The *Journal* and the *Herald* always blamed the unsolved crimes, and any crimes by Irishmen, on the ubiquitous Mollies, whoever they were, so I didn't connect McKenna and the Pinkertons with any of it. Also, we now know that the Pinkertons prevented Bully Bill from revealing publicly that it was McKenna's gang who had shot him.[18]

While he was bolstered up in bed covered with patches and plasters the day after the shooting, Bully was visited by Captain Linden. The big policeman forcefully urged him to promote the story that there was so much smoke and confusion in the stable that he couldn't make out the faces or figures of his attackers. This was a lie, because Bully had clearly seen the face of the man in the white slicker, and also the face of the frightened youngster who had shot him through the neck. Nevertheless, as he later testified, Linden instructed him to lie and so he did.

Linden explained to Bully that his police were out trying to discover who had done the deed, and the perpetrators would be less

---

18 Pinkerton, pp. 337-340; Comm. v. Kehoe, re: Wm. Thomas, pg.109

likely to flee if they believed Bully couldn't identify them. This, of course, was a lie, because Linden already knew full well who had done the deed. His own drinking buddies.

The Pinkerton Agency's reason for hushing up the affair seems obvious. If two or more of McKenna's gang should be arrested, then even if McKenna himself were not implicated, there would be public exposure of the gang and recrimination among its members. McKenna would risk losing his ability to lead them in tandem with Linden, as he had become accustomed, and that could not be permitted. There were many things yet to be completed on their agenda of death.

I blissfully continued about my own business and the study of law, only vaguely aware that the legal system and public opinion were being grotesquely fashioned to fit the designs of demagogues.

*   *   *

For those not familiar with the historical record, it should be noted that of all the men allegedly involved in the conspiracy to murder Bully Bill, only Kehoe and the A.O.H. leaders who attended the June 1st meeting, and the two young sacrificial goats, Gibbons and Morris, paid any legal penalty.

McKenna was not only exonerated because he was a detective, but he was treated as the hero of the day. His technical Bodymaster, McAndrew, received a pardon for all of his crimes as a reward for loyalty to McKenna. McKenna helped his henchmen, Tom Hurley and Mickey Doyle, in making good their escape. They were never put on trial.

Yet this was sufficient to slake the public thirst for vengeance. It was managed to be directed against the Molly Maguires as opposed to the individuals who had done the deed.

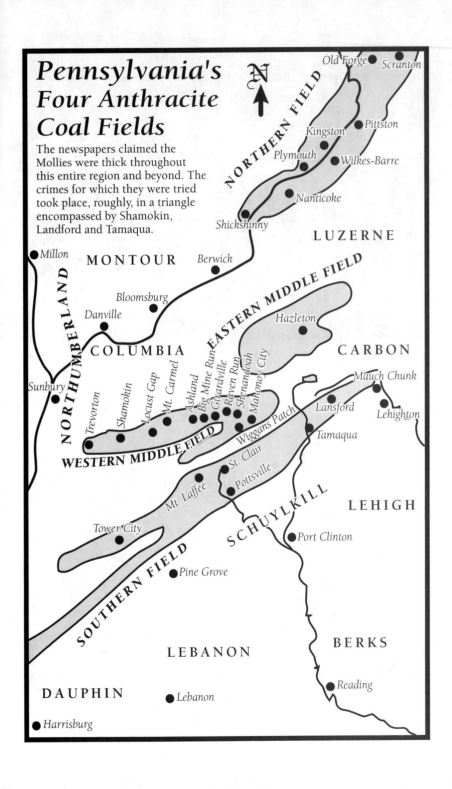

Pennsylvania's Four Anthracite Coal Fields

The newspapers claimed the Mollies were thick throughout this entire region and beyond. The crimes for which they were tried took place, roughly, in a triangle encompassed by Shamokin, Landford and Tamaqua.

N

NORTHERN FIELD

Old Forge
Scranton
Pittston
Kingston
Plymouth
Wilkes-Barre
Nanticoke
Shickshinny

LUZERNE

Millon
MONTOUR
Berwick

Bloomsburg
Danville

EASTERN MIDDLE FIELD

Hazleton

CARBON

COLUMBIA

Sunbury

NORTHUMBERLAND

Mauch Chunk
Lansford
Lehighton

Trevorton
Shamokin
Locust Gap
Mt. Carmel
Ashland
Big Mine Run
Girardville
Raven Run
Shenandoah
Mahonoy City
Wiggans Patch

Tamaqua

WESTERN MIDDLE FIELD

St. Clair
Pottsville

Mt. Laffee

LEHIGH

Tower City

SCHUYLKILL

Port Clinton

SOUTHERN FIELD

Pine Grove

BERKS

LEBANON

Reading

DAUPHIN

Lebanon

Harrisburg

IRISH EMIGRATION

As mocked by the English press.

As witnessed in Ireland

The Enterprise Breaker

An unidentified colliery operated by the Philadelphia & Reading Coal & Iron Company. On left Capt. Robert Linden of the P&R C&I Police and George F. Baer, president of the P&R C&I Co. Courtesy of Shamokin News-Item.

THE DRIVER BOY

The Driver Boy controlled mules which pulled coal cars over rails throughout the mine. He was usually 14 or 15 years of age. He has the oil-burning lamp, rubber boots because the driver walked in the ditch, and a braided leather whip called a "black snake." This is how one of them looked in 1890.

Breaker Boys picking slate in 1885 at Eagle Hill Colliery, P&R C&I Co. New Philadelphia, PA. The slate picker boss always carried a stick and knew how to use it. Courtesy of Shamokin News-Item.

Mule drivers, door boys, and spraggers at the close of a day in a mine, 1915.

Carbon County Jail

Mule drivers in a typical anthracite patch town.

Above is a typical coal "patch", the dwellings of miners in the early days of the Anthracite regions whose remains are still evident in certain sections. Courtesy of Shamokin News-Item.

Allan Pinkerton, owner of Pinkerton's National Detective Agency. He allegedly thwarted a plot to murder President-Elect Lincoln and acted as Chief Intelligence Officer for Federal General G.B. McClellan during the Civil War.

Captain Robert J. Linden, the leader of "flying squads" of Coal & Iron Police and also a Pinkerton Detective.

"Dan, show us your coat!" said Kehoe. Pinkerton's depiction of Dan Dougherty complaining to the "Mollies" that he'd been shot at after his acquittal. Note the pictures of the crucifixion and the Pope on the walls of Clark's saloon.

Franklin B. Gowen

James McParlan

Alex Campbell

John Kehoe

Jimmy "Powderkeg" Kerrigan

STRICTLY CONFIDENTIAL

The following are FACTS for the consideration of the Vigilance Committee of the Anthracite Coal Region, and all other good citizens who desire to preserve law and order in their midst. viz: —

On June 28th, 1875, at about 5 A.M., an attempt was made to murder William Thomas, otherwise known a "Bully Bill," by John Gibbons and Thomas Hurley. Mr. Thomas was shot at and wounded by these men as he was quietly conversing with William Heilner, Stable boss of Shoemakers Colliery, and was entirely unaware of his danger. —John Morris and Michael Boyle were accessories to the shooting, and present when it was done.

On July 5th, 1875, at 2 A.M., Police Officer Frank B. Yost, of Tamaqua, was shot and killed by Hugh McGehan as he, (Yost) was turning off the Gas of a Street Lamp a few rods from his residence. —James Boyle was present, and accessory to this murder, which was instigated by James Kerrigan.

On August 14th, 1875, at about 9 P.M. THOMAS Gwyther, a Justice of the Peace of Girardville, while standing at his office door in company with his daughter, was shot dead by William Love.

On the same date at about 11 P.M., Gomer James, a watchman, was shot and instantly killed at a Pic-nic at Glovers Hill; Shenandoah, by Thomas Hurley.

On September 1st, 1875, at about 7 A.M., Thomas Sanger, a Mining Boss, and William Uren, a Miner of Raven Run, were shot and fatally wounded by James O'Donnell, alias "Friday," and Thomas Munley, as the unsuspecting victims were on their way to their work. — Charles O'Donnell, Charles McAllister, and Mike Doyle were present, and accessories to this murder.

On September 3rd, 1875, at 7 o'clock, a.m., John P. Jones, a mining boss at Summit Hill, while on his way to work, was shot dead by Michael Boyle and Edward Kelly. This murder was instigated by James Kerrigan, who was accessory to the same.

. . . .

RESIDENCES OF MURDERERS AND ACCESSORIES: —

John Gibbons,	Member Molly Maguires'				Shenandoah.
Thomas Hurley,	"	"	"		"
John Morris,	"	"	"		"
Michael Doyle,	"	"	"		"
Hugh McGehan,	"	"	"		Summit Hill.
James Boyle	"	"	"		"
William Love.	"	"	"		Girardville.
James O'Donnell,	"	"	"		Weigans Patch
Charles O'Donnell,	"	"	"		"
Charles McAlister,	"	"	"		"
Thomas Munley,	"	"	"		Gilberton.
Michael Doyle,	Secretary				Mount Laffee.
Edward Kelly,	Member				"
James Kerrigan,	Ex-Body Master				Tamaqua.

*Wormley Hotel catered to the
rich and powerful
– Washington's political elite in the 1870s.*
Courtesy of Washingtoniana Collection

CHAPTER 11

KERRIGAN MURDERS OFFICER YOST

While the plot to murder Bully was being nurtured beyond my knowledge, I watched the end of the strike and the disintegration of the union with dismay. In spite of our dances, picnics and speeches, and our wonderful march of solidarity, the operators didn't budge from their demands for pay cuts. They kept opening the mines with blacklegs. On June 13 Captain Linden, headquartered in Shenandoah, reported that the strike was about ended. On June 15 the Miners' and Laborers' Benevolent Association of Northumberland County called a meeting to ask the operators to take them back at a ten percent reduction instead of twenty percent, but the operators distained even to show up.[1] On June 16 the Court of Clearfield County issued fines and jail sentences to the men who had protested the use of blacklegs there, seeming to justify the widely held belief that union activity was criminal activity.[2] The W.B.A. had been ineffective, its funds were exhausted, its membership drained away, and it died. I was let go in the first week of July 1875.

For several weeks after the death of the W.B.A. and the end of the strike, it might have seemed to the ignorant that peace had come to the coal regions. The miners and the mechanics and the laborers all went back to work, and the merchants' bills were paid, and there was no more marching, and no more defiant arguments in the taverns or

---

1 *MJ*-June 15, 1875
2 *MJ*-June 16, 1875

unusual clusters of men on the street corners. The workingmen had been crushed by their need for work, even at the lower rates, and there wasn't a damned thing they could do about it.

It was only after the collieries had been operating for a while that the union began to be missed as the salve for soothing the inevitable scrapes between those returning to work and those who enforced the operators' rules, who decided which breasts to allocate to whom, what safety conditions were adequate, what was a proper payload and what was too much gob, and what would be charged for materials used, and which men would be permitted to return, and who would not. In a work environment more dangerous than the front line of a battlefield, any of these issues could come to be regarded as a matter of life or death. For a short time, before these sores festered and became virulent, there was little labor-management violence.

But on July 5, 1875, there was a murder that outraged the entire coal region. It seemed at first to have nothing to do with the mining business or even race relations. Even with hindsight, I don't think that at its inception, it had anything to do with McKenna and his gang. Their *modus operandi*, as displayed in the assault on Bully Bill, was to commit daylight shootings before a multitude of witnesses, almost in contempt of being caught. This job was done at night by men who made an effort to conceal their presence at the scene. It was more like the fabled Molly Maguire operation described by McKenna in his early reports – a job known only to those with a need to know, those who planned it and carried it out.

It was conceived in the hate-twisted brain of a conniving ne'er-do-well named Jimmy "Powderkeg" Kerrigan. He was short and slight, not over five-feet or a hundred-and-twenty pounds. But using lies and trickery, sheer meanness, hair-trigger viciousness, and the ability to nurse a grudge forever, he had elevated himself to be absolutely the worst character in the borough of Tamaqua, where he lived.[3]

---

3 See: *Pottsville Evening Chronicle-PEC*-March 27, 1876.

He feared neither God nor man – no one except his wife, a decent woman who had borne him four children and who was more than double his size and strength. Though he sometimes worked, he hung out in the bar of the Union Hotel in Tamaqua, which, everyone suspected, was his headquarters for planning occasional acts of thievery, robbery and burglary. He was the local A.O.H. Bodymaster, put in place by Kehoe's predecessor, the mentor of McKenna. Everyone knew him and everyone with an ounce of good sense avoided him.

Unfortunately for the policemen of the borough, they were not always able to avoid Kerrigan. The two that made up the night patrol lived within five-hundred yards of him and their usual rounds took them past the Union Hotel where he and his cohorts hung out. They were Frank Yost and Barney McCarron, an efficient and reliable German-Irish partnership that was well respected and genuinely liked by all decent citizens. Their commission to keep the peace had often required them to put an end to the drunkenness and brawling of Kerrigan and his chums. Yost had drummed his nightstick on the diminutive Bodymaster's head more than once and, more seriously, McCarron had held him back from fighting with a fellow and in the process the opponent had slashed Kerrigan's face with a knife. It was common knowledge in Tamaqua that Kerrigan and his pals were seething to revenge themselves upon the policemen.

During the Independence Day festivities that year, observed on the 5th because the 4th was a Sunday, officer McCarron was freshly warned that Kerrigan and the gang from the Union Hotel had been overheard talking loudly about putting him "down." He took the warning seriously. He was the most vulnerable when, early in the morning, he had to climb upon a stepladder and put out the street lamps. So at his suggestion, his buddy Yost went with him on his rounds, and he went with Yost. The night was cloud bound and dark.

The partners had a snack in officer Yost's kitchen at two o'clock in the morning before going out to extinguish the last of the lamps. From an upstairs window Mrs. Yost sleepily watched her thirty-three-year-old husband cross the street to the corner nearest their home and ascend his ladder under the lamppost. Abruptly the night was pierced by the flash and retort of two shots and her man fell.

" Oh, God! No!" Mrs. Yost screamed and hurriedly put on a dress while running to his assistance.

Barney McCarron was on the opposite side of the street walking away from his partner when he heard the shots. He turned, drawing his pistol. He saw Yost staggering back toward his house calling for his wife, and ran past him in pursuit of the two assassins. They ran from him and away from the light of the street lamps, into the town cemetery. It was dark there and the policeman could no longer see them but he could hear their footfalls, and in that direction he fired two shots. There was one answering shot from the cemetery, and then silence. McCarron saw Yost in the arms of his wife and he dashed the 120 yards to the house of the doctor for the neighborhood.

The neighbors helped the doctor take Yost into his living room, and after an examination it was clear to all that his wounds were mortal. He had seen the men who had shot him, by the light of the lamps. One was a big fellow and the other was smaller. He didn't know their names, but he had seen them earlier that night in company with the Irish gang from the Union Hotel.

The doctor asked him point blank if it were not Powderkeg Kerrigan and one of his pals who had shot him, and he answered no, the two were strangers from over the mountain.

He said he had the impression that the assassins had really been trying to get his partner Barney, and while his back was turned to them in his uniform, they had shot him by mistake.

But the dying policeman had no doubt whatsoever about who was responsible for his death, and more than once he warned his distraught partner, "Barney, beware of Kerrigan or they'll get you yet, the same as they have me!" [4]

---

4 This version of events may be verified in the *Pottsville Evening Chronicle,* **May** 10 -15, 1876.

He died between nine and ten o'clock that morning, and his murder sent the coal region into an uproar. His funeral was so well attended that business in the entire borough of Tamaqua was suspended that afternoon. Leading citizens, led by his wife's brother-in-law, formed a committee to marshal evidence against Kerrigan and his gang. The committee hired the Pinkerton Agency, which confided to them that it already had an operative in place who would be assigned the task. The outcry in the newspapers against the murders was unanimous in demanding that swift vengeance be meted out remorselessly to the Irish under the code name of the Molly Maguires. Even though Yost's partner – and in Yost's mind the intended victim – was an Irishman, the attitude of the other races hardened against them. They were explicitly given to understand that their reputation for rampant individualism, belligerence, disrespect for authority and lack of discipline were traits no longer to be tolerated. The dreaded spectre of the murderous Molly came to dominate the public mind, displacing the image of the faithful and grieving partner Barney. Good jobs became ever more difficult to find for the Irish and, of course, resentments festered.

CHAPTER 12

McKenna "Turns" Kerrigan

I'm often amazed by the irresistible need of some criminals to talk about their crimes. One of my clients brutally shotgunned a District Judge, creating widespread public outrage, and yet two days later he told what he'd done to a group of guests attending a birthday party! Every defense lawyer has a similar story.

Kerrigan needed to let people know that he had taken vengeance upon Yost. He visited Bodymaster Roarity of Coaldale on the 14th or 15th of July[1] to assure him, and the two assassins supplied by him, that all was well. He couldn't help bragging about having pulled off a very "clean" job. He only told his friends. But friends tell their own friends, and so forth, until the information is received by some who have insufficient incentive to be careful with it.

On July 15, detective McKenna boarded the train to Tamaqua for the purpose of investigating the murder of officer Yost. His method of investigation is revealed by the entries in his expense account. He got off the train to visit two barrooms where he spent a whopping $1.70 on "treats" for those he met. The word "treats" was his euphemism for drinks. That $1.70 was, at the time and place, the price of more than seven gallons of beer – or whiskey of even greater alcoholic content.

He continued on to Tamaqua where he wisely paused to have a

---

1 *Evening Chronicle*-May 12, 1876, testimony of J. Kerrigan.

50-cent dinner. But then he played the jolly host at Carroll's Union House (85 cents) and at Nolan's bar across the street (45 cents) before taking the cars further east to the Lansford area where he "treated" to the tune of $3.85 (16 gallons of beer) at three other saloons.

McKenna's expenditures for drinks continued at a similar rate throughout the next three days! One might say that he sloshed into his investigation on a wave of alcohol. Certainly, he came out of it with a hangover.

This isn't to imply that his methods weren't effective. They were. He drank a lot, but most of the time he managed to keep sufficient of his wits about him to understand and remember the drunken blathering of those he "treated". He didn't have to ask questions because sooner or later the conversation turned to the hottest topic of the day, the Yost murder. He was freely given to share the "inside story" from men in a position to know. This kind of hearsay information, although not useable in court, was sufficient to his immediate purposes.

He learned that indeed, as everyone in Tamaqua suspected, Powderkeg Kerrigan had been the prime mover behind the murder. Sick and tired of being humiliated by Yost every time he got drunk, he had more than once threatened to kill the policeman.[2] He knew he would be suspected, and that should anyone see him at the scene of the crime he would be easy to identify because of his unusually small size. So the night of June 5th he went over the mountain to Coaldale near the Schuylkill/Carbon County line and brought back two men who would do the actual shooting. The men had been recruited for him by the A.O.H. Bodymaster in Coaldale, named Roarity.[3]

Kerrigan took care to keep the two young men out-of-sight inside his hangout, Carroll's Union House, except when he took them

---

2 *Evening Chronicle*-May 15,1876, test. of Patrick Duffy.
3 *Evening Chronicle*-May 15,1876, test. of Kerrigan.

outside and identified Yost to them as the policeman walked his rounds. One of them didn't have a suitable pistol. Kerrigan went to several places trying to borrow one before settling for a single shooter from his friend, the proprietor of the Union House, Jim Carroll. At about eleven o'clock p.m., with all in readiness, he pretended to go home for the night and left the Union House. But on the way, relishing his evil taste for revenge, he sought out his intended victim, Officer Yost, and cynically offered to buy him a drink. He did go home then, but he stayed only long enough to get a bottle of whiskey. He sneaked back out and without being seen, made his way into the Tamaqua Odd Fellows Cemetery. There, as planned, he met the two who would do the killing, and all three drank whiskey and waited in the dark for Yost to bring his ladder under the street lamp.

Kerrigan stayed crouched behind a tombstone while the assassin with the revolver pumped two bullets into Yost. Then he hustled the young men through the maze of the pitch-dark cemetery and the less frequented alleys of the town, without being seen by anyone. He put them on the road back over the mountain and furtively returned, again without being noticed, to his own front porch. There he finished his bottle, confident and elated that he had pulled off a clean job; a murder for which he could never be convicted.[4] He tried to enter his front door and found it locked, so he rattled the doorknob until his wife woke up and let him in. They had a conversation and he told her that he had shot Yost, but if she ever spoke of it, he would blow her damned brains out.[5]

* * *

This basic story, excluding the words between husband and wife, of course, was pieced together by McKenna from various people over the course of his four-day bacchanal. There were no surprises in it.

---

4 This account substantially conforms with Kerrigan's testimony at trial, except it omits his absurd claims that everything he did was instigated by Carroll, proprietor of the Union House, and by his own feckless but favorite drinking companion Duffy.
5 *Evening Chronicle*-May 17, 1876, test. of Mrs. Kerrigan.

Except for some interesting details, it was just about what McKenna had expected to learn before he had got on the train to leave Shenandoah.

But on the third day of his carousal, McKenna stumbled upon a connection so surprising and so important to the designs of Railroad President Gowen that he cut short his trip and started on his way back to Shenandoah the next morning. When he got to his boarding house he penned a lengthy report to the Pinkerton Agency's Superintendent in Philadelphia.[6] Once received by the Agency, that report has never again been brought into the light of day. I doubt if even Mr. Gowen ever saw it. I know what was in it only because the Pinkertons admitted some of its contents,[7] and the rest may be inferred from what we know they did with the information.

What McKenna discovered which was so important to his Agency was a name – the name of the man with the revolver who actually shot Yost. The other one with the single shooter, who never pulled the trigger, was a usually inoffensive fellow who had done what he did largely because he didn't have the gumption to say "no" to those whom he mistook to be his betters. But the actual shooter was identified to be none other than Hugh McGeehan.

Hugh McGeehan had made a name for himself as one of the rising stars of the labor movement which Mr. Gowen was anxious to crush. Although young, in his twenty-third year, he was virile and handsome and a fine public speaker. In the late troubles, he had been the outspoken leader of many a workingman's demonstration. But there was a price to be paid for such leadership. After the strike when the mines reopened, he was blacklisted.[8] He found there was no work for him in any of the collieries of Schuylkill or Carbon Counties. This made him desperate to earn a living, particularly since he was engaged to be married. He began to nurture a deep hatred for the boss responsible for the blacklisting, one John P. Jones, the

6 Pinkerton, pg. 387; *Evening Chronicle*-May 6, 1876, test. J. McParlan.
7 Testimony of McParlan in the first Yost trial.
8 *Evening Chronicle*, March 4, 1876; testimony of Wm. Evans, May 15,1876.

outside superintendent of collieries nos. 4, 5 and 6. When Kerrigan came looking for someone to shoot Yost, McGeehan agreed to do it in exchange for Kerrigan's promise to kill John P. Jones.[9]

All that aside, what made the identification of McGeehan so surprising and important was that the young man happened to be one of the chief lieutenants of Carbon County's most dynamic labor leader, Alexander Campbell. Using the A.O.H. as the organizational framework for his activities as a labor leader, Alex Campbell was cast in the same mold as that other A.O.H. leader from Schuylkill County, Jack Kehoe. Like Kehoe, he insisted upon strict discipline from the rank and file of his workers and made them disavow all forms of violence. As a result, the Pinkertons and the mine operators and the pro-violence faction within the A.O.H. detested Campbell as much as they detested Kehoe. Therefore, it was important and surprising that one of Campbell's lieutenants could be convicted of murdering a man in exchange for a murder to be done in the future. McKenna's mouth must have watered at the prospect. He had spent over a year surrounding Jack Kehoe with violent acts and trying to make some of them stick to him, and now here was Campbell being served up to him, already connected with the Yost murder and with another one to take place soon!

I do believe that Alex Campbell and Jack Kehoe were both targets of the Pinkerton/Gowen conspiracy before McKenna ever came into the coalfields. Both of them had long been at odds within the A.O.H. with the Pinkerton Agency's informant from that order, Barney Dolan,[10] who as County Delegate had promoted McKenna to his position of power in the Shenandoah Division. They both had fought Barney Dolan when he had promoted Powderkeg Kerrigan to a similar position in the Tamaqua Division[11] and they combined to deprive Dolan of his chair as County Delegate.

And a final fact must be noted as enhancing the Pinkerton

---

9 *Evening Chronicle*-May 12, 1876, test. of J. Kerrigan, page four.
10 See Author's Notes.
11 *Evening Chronicle*-May 12, 1876, test. of J. Kerrigan, page four.

Agency's ability to exacerbate trouble from all this. Both Kehoe and Campbell happened to be related through marriage to the O'Donnell clan of Wiggans', Patch, whose reputation for fighting with the Modocs and threatening the mine bosses with coffin notices was a persistent source of embarrassment to them both.

McKenna realized that the situation was uniquely portentous. In his report to the Agency he suggested that he and Captain Linden should meet with their superiors from the home office to discuss how they might make the most of this opportunity for mischief.

*   *   *

A few weeks later, I was lodged in a Tamaqua Hotel on the business of interviewing witnesses for the Ashland attorney out of whose office I was reading the law. During those weeks detective McKenna had managed to entice Kerrigan and his boys at the Union House to brag to him about their clean job. These direct admissions had eliminated his legal problems over hearsay evidence. Also, the Pinkerton Agency, in response to McKenna's suggestion, had treated him and Linden to a short vacation at the Glen Onoko Summer Resort. There, the field agents with agents from the Philadelphia office had elaborately laid out their plans for ensnaring the targeted A.O.H. leaders. They had probably consulted with Mr. Gowen.[12]

When I saw McKenna at my hotel in Tamaqua, I assumed he was there to investigate the Yost murder. I knew, as did everyone, that Kerrigan was widely suspected of being behind the murder. If McKenna was there to convict that little ruffian, I had no objection.

I avoided contact with the detective. His raucous habits weren't compatible with the quiet interviews I needed to conduct. I'd arranged for exclusive use of a small cardroom off the back of the bar, where my witnesses came to meet me after the day-shift had been let go.

By the time I had finished interviewing the last of them, it was

---

12 Editor's Note. See: W. G. Broehl, *The Molly Maguires*, pg. 227.

almost dark. In the barroom, McKenna was nowhere to be seen. In search of an evening breeze as relief from the summer heat, I stepped outside of the hotel. I was almost run over by Captain Linden charging heedlessly down the boardwalk. I knew him but he didn't know me.

"Watch yourself, there!" he ordered as he swerved to get around me and then resumed his course like an inexorable, blue-clad, brass-trimmed juggernaught. He was carrying a small lantern.

I don't know what came over me then, but as his broad back swaggered into the gathering dark, I followed him. I needed to reassure myself about what the Pinkertons were doing with the secret I had permitted them to keep. Still, I wasn't accustomed to sneaking about, and I felt a sense of shame as I followed him.

He proceeded directly to the most remote section of the town cemetery and seated himself on a small gravestone, placing the unlit lantern on the grass between his feet. There were other, larger monuments in that area, and when full darkness enveloped him the big cop's bulk became indistinguishable from them.

A significant time passed, perhaps half an hour, which I occupied by inching myself as close to the policeman as I dared.

Finally, I heard men's voices coming along the wagon path through the graveyard. They were not guarded voices and I could make out one of them to be McKenna.

"I must admit, Powderkeg, that this is as good a place as we could pick for holding a private conversation."

To another man he said, "Ye can leave us now, Edward. See ye back at the hotel."

The other man walked away, and though he was only a silhouette in the starlight, I thought I recognized the singularly odd shape of the bodyguard I knew only as Whitey.

When he had gone, Kerrigan complained about him, "Ye ought to make that one develop a sense o' humor. I tried to kid wid him that I was too busy to meet ye here and he almost broke my arm. He's got a grip on him like an eagle's talon!"

"Is that so?" chuckled McKenna. "Yes, I suppose he does tend to be too intense at times. Sorry about that, Jimmy."

McKenna sat on a bench along the pathway, and joining him Kerrigan demanded, "So then what's so all-fired important?"

"It's a matter of life or death, Jimmy . . . yer own life or death!" McKenna answered ominously.

"What the hell d'ye mean by that?!" Kerrigan exclaimed.

"Listen, Jimmy, I'm trying to help ye," McKenna said soothingly. "If all those involved in the Yost murder are going to hang anyway, what harm is there in saving yer own neck by cooperating with the police?"

"That's nonsense, McKenna!" the little Bodymaster cried. "Ye know full well it was a very clean job!"

"It would have been, Jimmy, except that you an' yer buttys confessed everything to an undercover operative of the Pinkerton Detective Agency."

"What?! Who?! No!" Kerrigan exclaimed and jumped in panic to his feet.

There was a flash of light as Captain Linden struck a sulphur match, brought up his lamp and bellowed, "Sit down, Kerrigan, you stupid shit! Sit down and listen!"

Kerrigan sat as though obeying an order from the occupant of one of the graves.

"Why are you being so nice to him, McParlan?" growled the Coal & Iron Captain as he advanced menacingly upon the men. "He can either do what we tell him or he can be the first to hang!"

"I think we should try to have a calm discussion, Captain," said the undercover agent putting his hand reassuringly upon Kerrigan's shoulder to introduce him, "This here is Captain Linden of the Coal . . ."

"I know that!" interrupted Kerrigan, brushing away the comforting hand and standing again. "But who an' what in the hell . . .?!"

The detective smiled and said, "Ah yes, Jimmy, my real name is James McParlan. Ever since ye've known me I've been an operative of Pinkerton's National Detective Agency. Only doin' me job, ye know."

"I'll be damned!" cried Kerrigan.

"Ye probably are," joked the detective. "Therefore, ye'd be wise to try an' stay alive as long as possible."

This was sobering to Kerrigan. He sat again on the bench holding his head in his hands silently. Finally, he groaned, "I don't want to be an informer!"

"You've already informed, Jimmy," argued McKenna. "Ye can't change that fact. Not even if ye hang fer trying."

"And make no mistake about us, Kerrigan," enjoined Captain Linden. "We want much more from you than the information we already have."

"What is it you do want?" pleaded Kerrigan. "Exactly what?"

"We want you to testify," answered Linden. "But that isn't the most important. We want you to personally lead two men in an attack on John P. Jones. We want you to do it so poorly as to make certain that you're all caught in the act."

"So the Pinkerton Agency can take credit for the capture!" Kerrigan suggested derisively.

"Not at all!" answered Linden. "Except for the heroic exploits of our lone undercover agent here, we want to downplay the role of the National Agency. We want it to be one man against the Mollies."

"Then who's to catch us?" queried Kerrigan.

"I've organized Vigilance Committees in most of the towns," explained Linden. "We'll arrange to have you captured by one of them."

"A Vigilance Committee might be inclined to hang me on the spot," objected Kerrigan.

Linden laughed at this, but he reassured, "Don't worry, Kerrigan. What we want out of all this are successful prosecutions. We can't have trials without defendants. We'll let the Vigilante leaders know that you're working for us."

"You seem to have thought of everything – all planned out," said Kerrigan.

McKenna answered, "We do for the most part, but there are still some details not yet settled. We won't act on this further until after the holding of the A.O.H. Convention at the end of the month. That'll be an opportunity for us to ensnare some more of them."

Linden added, "The three of us will have to meet a few more times between now and then. The only issue now is whether you choose to cooperate or to hang."

"Will I have to spend time in jail?" asked Kerrigan.

"During the trials, jail will be the only safe place to keep you," answered Linden. "Over the course of those trials the time in which the Commonwealth must try you under its two-term rule will be permitted to expire."

"So you're saying I'll be let go?!" exclaimed Kerrigan. "Would the District Attorney ever agree to such a thing?"

"He already has," answered Linden. "He said that once you do what we ask, you can go west and raise grasshoppers, for all he cares!"[13]

After a long pause, Kerrigan said resignedly, "There really isn't any choice. None of my buttys would turn down such an offer."

"You're right about that, Jimmy," said McKenna.

"Good, that's settled then!" Captain Linden announced his blessing. Then he put his lantern up to his own face to illuminate a threat.

"Back in '68," he recounted, "one of the Reno Gang made a similar deal with Mr. Pinkerton, but afterward reneged on it. The gang was captured anyway and lodged in a Missouri jail. One night some highly disciplined Vigilantes broke in and hanged them all. The toes of the one who had reneged were permitted to touch the floor, just enough so it took him an hour to die!"

He lowered his lamp and he and McKenna walked away without another word spoken. Kerrigan stood in the dark a few minutes, in a state of shock. Then a night bird cried and the little man bolted and

---

13 *Evening Chronicle*-May 12, 1876, test. of J.Kerrigan, cross exam.

ran as though pursued by the fiend of hell.

*   *   *

It is here that my story substantially veers from what I can verify from Pinkerton's book or the detective reports. The Pinkertons claim that Kerrigan didn't start working for their organization until after he was in jail for the murder of John P. Jones.[14] I claim that he had already been recruited by the Pinkertons and was working for them when he led the two men to murder Jones.

Should the reader believe me, or the Pinkertons? McKenna's and Linden's reports admit that prior to the Jones murder, they tried for weeks to entice Kerrigan into the Tamaqua Cemetery to recruit him. They claim they couldn't get the little man to meet them there.[15] Of course, Kerrigan would have shown up in hell for the promise of a free drink. Such unbelievable claims support my version of what happened. As I'll later point out, Kerrigan's and the Pinkertons' subsequent actions in the assault on Jones also support my version. Here, I'll address only the issue of credibility.

Even at this late date, my own unsullied reputation after a lifetime of lawyering must stand for something. Conversely, we have already seen that McKenna was a liar, even under oath. We will now examine McKenna's field reports to reveal that he deliberately falsified them in order to implicate those he had targeted to be convicted of crime. Thus, his reports falsely attempt to implicate John Kehoe in his own gang's long-standing plot to murder the Welsh pistolero, Gomer James.

McKenna's report for July 4, 1875 claims that he and some henchmen went to Kehoe's in Girardville to get men to murder James. "When . . . the operative informed Kehoe what was the object of their visit to Girardville he said they could not get a man there for the job who was worth anything, but thought perhaps they could at

---

14 Pinkerton-pp. 464-468.
15 Linden Report of Aug. 22, 1875; and McKenna report of Aug. 22, 1875 in Kaercher MSS.

Big Mine Run." They went to Homerville but the Bodymaster of that place was not at home. They returned to Kehoe's where they asked the Girardville Bodymaster to supply the assassins, but he refused and angry words were exchanged. Finally, still in Kehoe's bar, they were promised the men by an old fellow who was County Secretary of the A.O.H.[16] So says the report.

I was shocked to the core when I read this. Could it be true that McKenna and his men had spent most of the day in Kehoe's Hibernian House in Girardville asking for men to murder Gomer James? That couldn't have happened unless Kehoe was at least complicit in the foul deed. My disposition to regard Kehoe as a good man began to crumble.

Then I compared the report of McKenna's activities with the Agency's expense ledger for McKenna on the same dates. The ledger is very specific about where and for what the money was spent. It does not support McKenna's claim that on July 4, he went to Kehoe's in Girardville. To the contrary, it proves that he spent the entire day at Big Mine Run! The entries for July 4 are as follows:

> July 4: *Share of buggy hire to Big Mine Run* $2.00
>   *Treating Donohue at crowd of MM's* $1.25
>   "   *Frank McHughes and crowd of MM's* $ .85
>   "   *Crowd of MM's at Big Mine Run* $1.25 [17]

Thus, McKenna's own expense ledger reveals that he was a liar – indeed he was the deadliest liar in the annals of American Jurisprudence. It may also be noted that he was spending enough money on booze to embalm the livers of the Light Brigade. This evil persistence of the Pinkertons to entrap Kehoe makes me curse myself

---

16 McKenna's report summary for July 3 and 4 and expense reports for June 29-July 10, 1875 are set forth in their entirety in the Appendix.
17 McKenna's route to Big Mine Run would have gone through Girardville. But the issue is whether he spent such considerable time and made such considerable contacts in Kehoe's Hibernian House. He didn't, or his expense account would have noted it. Big Mine Run was the lair of Barney Dolan, McKenna's mentor in the A.O.H. who had put him in his position of authority in Shenandoah. Dolan hated Kehoe who had ousted him as A.O.H. County delegate.

for not telling him that McKenna was a detective.

I suspect that McKenna got men to murder Gomer James on July 4 at Big Mine Run, and wrote up a false report so he could later claim that Kehoe had ordered him to murder James.

But the plan was fractured because Gomer failed to show up at the picnic where he was to be killed. About a month later, Tom Hurley did the shooting and was seen doing it by many witnesses. He had to go into hiding. It would be impossible to convict Kehoe of the deed. So, the Gomer James case was never brought to trial; the Pinkertons were not motivated to carefully launder the summaries and ledger relating to it.

In any event, the great detective certainly falsified his reports. Therefore, they should be relied upon only when verified by independent sources, or when containing declarations against the interests of the agency.

Before continuing with the story as I know it, there is a final observation to be made about McKenna's reports for July, 1875. Although the covering letter under which they were sent to Mr. Gowen expressly states that they cover the period from July 1 to July 24, 1875, the reports for the five days during which McKenna went to the Tamaqua area to investigate the Yost murder are all missing. The expense accounts for those five days, July 15 through July 19, are included in detail. Only the results of the detective's investigation are missing.

The question of why the notes of activities for those days are missing is a valid inquiry. Their absence casts doubt upon the reliability of the reports as a systemic recording of events at or near the time they occurred.

Much later, the Pinkertons supplied the District Attorney with a narrative for those days, to be used in obtaining convictions against those whom the Pinkertons accused of the murder of officer Yost and a related murder.[18] However, that the five days were not included in the original report summary to Mr. Gowen permits the inference that

18 Editor's Note. See:Kaercher MSS,George Keiser,Pottsville, Pa.

their original content differed from what was later given to the District Attorney.

I think its reasonable to believe that the narratives were altered by the Pinkertons – the better to convict some men who were innocent, and to exonerate some men who were guilty, including Powderkeg Kerrigan and McKenna himself.

CHAPTER 13

THE MURDER OF JOHN P. JONES et al

Before that night in the graveyard I had known nothing about Mr. John P. Jones – never heard of him. But knowing that the Pinkertons had not only discovered the plot against him but had taken control of the chief plotter, I assumed him to be the most adequately protected mine boss in the region. Sure, I did think the detectives had offered Kerrigan too favorable a deal by far, to be getting off with no time in prison. But it seemed they were diligently ferreting out real crime and not merely labor agitation. I was in favor of that. So I didn't run out and warn everyone about the detectives. The angel of death had not yet played lasting havoc with my youthful preference to find the good in everyone.

On August 5, 1875 the Superintendent of the Pinkerton office in Philadelphia mailed a formal report to its client, the Citizens' Committee of Tamaqua, which advised them there was a plot to murder Mr. Jones and suggested that they should warn him. None of their own operatives ever warned him. Also, Captain Linden verbally told one or more members of the Tamaqua Committee about the danger to Jones. He told them that the attack would be made during the day as Jones was going to or from work, and that Kerrigan was involved. The Tamaqua men in turn reported the plot to the Superintendent of the Lehigh Coal Co. who was Jones' boss, and it was he who talked to Jones.[1]

---

1 See testimony of B. Franklin, R. Linden, M. Beard and W. Zehner in Comm. v. Campbell

The Super gave Jones only a vague warning that he was in danger of an assault. He advised him not to go to work by walking along some old, abandoned railroad tracks as he sometimes did, and told him he should go up on the locomotive from Lansford every morning as the safest way.[2] Jones would do as he was told. He had been told by the Super to blackball McGeehan, and he had done it. The Pinkertons' plan did not call for the attack on Jones to be made in any out-of-the-way place where there would be no witnesses.

*   *   *

On the morning of August 9, 1875, Pinkerton operatives McKenna and Linden took separate trains to Mauch Chunk, the Seat of Carbon County, to meet there with the Superintendent of their Philadelphia office and to witness a ceremony. All three would testify that outside the County Recorder's office they saw McGeehan, the shooter of Yost, being rewarded by Alex Campbell. They would say that Campbell caused the County Recorder to issue McGeehan a license to operate a saloon as a reward for murdering Yost.[3] There is no doubt the license was presented in such a public way, but that it was a reward for killing Yost is nonsense.

It isn't possible to reward a man by paying him what is already owed to him. Well before the Yost shooting, Campbell owed McGeehan a debt for having promoted the union to the point of getting blackballed from mining. That debt pre-existed. Also, Campbell was a wholesale liquor dealer who sold his wares to the saloonkeepers, and helping the popular McGeehan to open a barroom was good business for both men. The arrangement was no secret. Rather, word of it was spread far and wide in order to drum up business for the new saloon. Thus, the public presentation ceremony. The Pinkertons' painting of the scene as one of sinister reward for foul murder was nonsense to anyone seeking the truth.

---

2 See testimony of W. Zehner in Comm. v. Campbell
3 See their testimony in Comm. v. Campbell

Unfortunately, it was believed by people preconditioned to believe in a widespread Molly Maguire conspiracy.

Captain Linden wrote a report of that day for the information of Mr. Gowen. He makes no mention of the liquor license ceremony. If he thought it was as important as he later testified, you'd think he would have mentioned it. He does report seeing Campbell that evening on the return trip, and he presents Gowen with a detailed description of him – " . . . age about 35 years, height 5-feet, eleven-inches, medium build, light complexion, brown hair, light moustache, sharp features – a raw-boned, muscular-looking man." [4]

Linden would not have bothered to provide a word picture of a man in whom the Railroad President would have no future interest, so it seems fair to infer that Campbell had been targeted for prosecution. But Gowen would never prosecute Campbell with respect to Yost. He would prosecute him for complicity in Kerrigan's attack on John P. Jones, which had not yet occurred. The inference is available that Gowen knew what had been planned for Jones.

* * *

On August 12, 1875, the Coal & Iron Police in Lansford were informed by the Tamaqua Committee that there would be an attempt on the life of John P. Jones. Their chief immediately warned Jones. He stationed a policeman in Jones' house to guard him at night, although the attack was planned to take place during the day and away from the house. This nocturnal police protection for the Jones house was diligently continued for three whole weeks, and was not discontinued until the night of September 1. That was the night when Powderkeg Kerrigan would arrive in Lansford with the two men who would make the attack. [5]

One might speculate whether Jones and his wife felt a sense of relief when the policemen were withdrawn from their home,

---

4 Linden report for Monday, August 9, 1875.
5 See testimony of Kerrigan in Comm. v. Campbell.

assuming it would not have been done unless the danger to Jones' life had abated. It seems likely that Jones felt somewhat safer when he walked alone to the Lansford Station to take the train to work on that morning of September 3. We can only speculate, because no one ever asked his widow that question.

*   *   *

During that three weeks when Mr. Jones' sleep was being so well tendered by the Coal & Iron Police, there occurred some very public acts of violence. Some were random acts that had little to do with any kind of organization, but they were cited by the newspapers to support their proclamation that there was a "regular system of murder" operating in the coal regions. Everyone knew this to be a reference to the alleged Molly Maguire conspiracy. The papers also declared that the organization of Vigilance Committees was the only option left for decent, law-abiding (meaning non-Irish) citizens.

Three of the acts of violence occurred on the same evening of August 14, 1875. At a picnic outside Shenandoah sponsored by the Rescue Hook & Ladder Fire Company, McKenna's gang finally consummated their months-old hunt for Gomer James. Those fire company picnics were grand affairs. I heard what happened from a musician from Locust Gap who was there as one of the band playing music at the dance pavilion.

He said that there were hundreds on the dance floor and, not far from it, around a tent where beverages were sold. It was a steamy hot night and cold beer was in great demand. Gomer James customarily worked this refreshment booth, as he did this evening. He had a remarkably well-muscled upper torso and he showed it off for the ladies by wearing nothing on it but a pair of red fireman's suspenders. As he bent and flexed to draw and serve containers of the frothing liquid, his muscles gleamed in the lamplight. He was well aware that the Cosgroves had put a price on his head, but he was defiantly confident that no one would dare attack him there in his own element.

He filled six heavy pitchers of beer. With three in each hand his biceps bulged as he turned to carry them to his admiring customers at the counter. It was then that Tom Hurley shoved through the crowd and shot three bullets into his chest, propelling him onto his back in a spray of beer and shattering glass.

Hurley had a man with him. The two leveled their pistols at the crowd and everyone – man, woman and child – dropped to the ground or stayed frozen where they were. The music stopped, and the dancers as well. Then Hurley and his companion, pointing their pistols at the slightest sound or motion, backed away and disappeared into the wooded recesses of the grove. It all took place within a few breathless moments.

"Has Tom Hurley been arrested?" I asked the musician.

"Bedad, no! He hasn't been identified to the authorities," he answered.

"Why not? There must have been dozens of people who recognized him."

"The bad blood between the Cosgrove and the Jones families could turn into a lasting feud," the musician explained. "No one wants to be seen as taking sides. They're keeping their mouths shut."

I knew that Tom Hurley was McKenna's man to the core, so when I heard it was he who had shot Jones, I began to suspect that things in Shenandoah might not be as I had imagined. Hurley would never have done the deed in defiance of McKenna. While it was possible that he'd earned the Cosgrove blood money without seeking permission from his chief, I decided that the next time I talked to Kehoe I'd tell him all I knew about the detective and ask what he thought.

Even the *Evening Chronicle* acknowledged that the murder of Gomer was in retaliation for the Cosgrove shooting.[6] But McKenna tried desperately to link the crime to Jack Kehoe. He came up with the story that Hurley made him go to Kehoe to get money from the

---

6 *Pottsville Evening Chronicle*, August 16, 1875.

A.O.H. for the shooting, and Kehoe referred the matter to an "inner circle" of Bodymasters, including McKenna, who eventually declared that Hurley was entitled to it. The only proof of Kehoe's involvement in this "reward" story is McKenna's testimony. Such a reward would have been contrary to the A.O.H. By-Laws. Also, the Cosgrove family had already posted a reward and had even increased its amount in negotiations with Hurley[7] who presumably collected it directly. There was no good reason for the A.O.H. to pay additional money. Any demand for it by Hurley would have been ridiculous. There is no evidence in the A.O.H. ledgers or anywhere that such an additional reward was ever paid to Hurley. Believing this story requires the payment of a slavish credence to the word of McKenna. Yet, there were those who paid it.

As I said, there was other violence that the newspapers added to the Gomer James murder in concocting their stew about a Molly Maguire conspiracy. Bully Bill Thomas and some Irishmen emptied their pistols at one another while rampaging through the streets of Mahanoy City, whose enlightened citizens had made a present of the weapon to Bill. In Girardville, fights broke out between Welsh and Irish gangs and a popular Justice of the Peace was shotgunned while trying to serve an arrest warrant.[8] Only the brave and firm efforts of Constable Jack Kehoe prevented more riots and bloodshed.[9] Two days later a saloonkeeper and some friends sitting on his front porch were shot up, though not fatally. One of Kehoe's brothers-in-law from Wiggans' Patch was eventually accused of the assault. The *Pottsville Evening Chronicle* mixed all of this together and concluded, "a regular system of murder is in existence."[10] A conspiracy theory was being established as "common knowledge."

*   *   *

---

7 Pinkerton, pages 298,299
8 *Pottsville Evening Chronicle*, August 16, 1875.
9 *MJ*, August 18, 1875.
10 *Pottsville Evening Chronicle*, August 17, 1875.

If there was, in fact, a regular system of murder in the coal region, it centered very suspiciously around the comings and goings of James McKenna.

On the morning of August 31, 1875, he and his henchmen, Tom Hurley and Mickey Doyle, met in Lawlor's Saloon in Shenandoah with three visitors – members of the O'Donnell clan who had come from Wiggans' Patch. The visitors had been promised the aid of McKenna's gang to murder a mine boss at Raven Run, two miles outside of Shenandoah. That much is admitted by everyone.[11]

Also not in dispute is the basic account of what happened the next morning.[12] The witnesses recall that five strangers were scattered among the miners who were waiting for the Raven Run mine whistle to call them to work. One of them wore the same long coat of McKenna's which Tom Hurley had worn in the assault on Bill Thomas, of a salt and pepper color, mostly salt, which is the color of white cloth exposed to coal dust.

The inside boss, Tom Sanger, left his house followed by a young Welshman who was boarding there. Sanger was rumored to be one of those bosses to whom "No Irish Need Apply." He had received several coffin notices.[13] As he neared the entrance to the breaker and shaft, he did not notice that he had been surrounded by the strangers. One of them stepped in front of him with a drawn pistol and fired. One fired from behind. Although Sanger was hit, he ran up a street to his right that led to the home of the owner of the mine. But another stranger had been posted in that street who fired two more shots into him. He ran through the door of one of the houses along the street and collapsed into the arms of the woman of the place, sending them both crashing to the floor. The young Welshman who was boarding with Sanger came running to his assistance, and was himself fatally shot. The five strangers fled along the road going north.

---

11 Pinkerton, pages 428 to 432; testimony of James McParland in Com.v. Munley, and elsewhere.
12 This account is taken from the testimony in Commonwealth vs. Munley, unless otherwise indicated.
13 MJ, September 2, 1875.

The owner of the colliery was taking breakfast when he heard the shooting. He grabbed his pistol and ran swiftly to catch up with the last two of the departing strangers. He exchanged gunshots with them before they entered the woods in the direction of Shenandoah. Then he went back and organized a posse of twenty men to follow them.

McKenna's story of his involvement in the murder is an artful dodge of responsibility.[14] He said he knew nothing about the visiting assassins until he awoke on the morning of their arrival. Only then did his roommate Mickey begin to tell him about them. Then Tom Hurley showed up. He also knew all about the assassins. The three of them went to breakfast in McKenna's boarding house and the two henchmen filled him in as to what was afoot. He says that Mickey Doyle borrowed his long whitish gray coat to wear on the assault, though the coat would have been six inches too long for the little fellow, and Mickey would not have been a stranger at Raven Run but would have been recognized by many, having recently worked there.

After breakfast the three of them went to Lawlor's Saloon where Hurley joked with Mickey that he had better shoot straight for the O'Donnell clan or they might shoot him. They left Lawlor's and walked to welcome two of the killers as they arrived in town. The third killer met them at Lawlor's a little later. After the meeting of the six men, the Wiggans' Patch boys, with Mickey Doyle as guide, left for Raven Run. McKenna swore that he wanted to warn Linden or someone about the impending murder but Tom Hurley stayed close to him all the rest of that day and, try as he might, he couldn't do it. It's obvious that he could have taken a moonlight stroll to warn the mine owner. I'm furious that no one asked him about that. Of course, he could have said he was afraid he was being watched.

Anyway, McKenna says he was in Lawlor's Saloon at 8:30 the next morning, again accompanied by Hurley, when the death squad

---

14 See testimony of McParland in Commonwealth v. Munley and Pinkerton at pages 428 to 432.

returned from Raven Run and reported to him their success. McKenna bought them drinks and bid goodby to the Wiggans' Patch men who departed. He helped the others to avoid the posse who came looking into all of Shenandoah's Irish bars, by taking them to a bowling alley.

This tale of McKenna's is full of transparent absurdities. He was the leader of the Shenandoah gang. It's inconceivable that he would not have known in advance that his gang had invited the Wiggans' Patch men to come and commit murder. Clearly, the breakfast gathering at McKenna's boarding house was to prepare to meet the visitors. If for some unaccountable reason the planned murder was being kept a secret from him, why would Doyle and Hurley reveal it to him when they did? Why would McKenna be allowed to meet with the killers if not to plan what they were going to do? It's fairly obvious that he knew about the affair from its inception, and had arranged to meet with the killers in Lawlor's Saloon the day prior to the murder. He himself was, very possibly, the shooter who had worn his white coat. McKenna's *modus operandi* was exactly – a morning assault while the victim was going to work. It seems more likely than not that he was the organizer and boss of the operation from start to finish. We may say, at the least, that his failure to warn the mine owner reveals an egregious lack of concern for human life.

*   *   *

The more I ponder the absurdities in the Pinkertons' story – especially the absurdity of an undercover agent in a hugely expensive investigation who is hired even though he refuses to testify in Court; and who totally controls his gang of thugs except when it's advantageous to the investigation that he not interfere with their murders – the more I feel stupid and inadequate for failing to sway the juries with them. Then I realize that the passage of time has dulled my memory of how successfully the press had spread the belief in the existence of a network of murderous Irish Catholic thugs.[15] After the Raven Run murders, the national newspapers also began to run the story. Without exception, they adopted the prejudice of their coal region cousins, concluding that the evils of the Molly Maguires "are carried on under a charter granted by the Ancient Order of Hibernians" and that "the purpose of the Molly Maguires, or A.O.H., is to kill people and burn down buildings."[16]

I suppose it's no wonder that when an honest miner testified at the trials that he had never personally known of or encountered a group called the Molly Maguires, he was laughed off the witness stand by the spectators and the jury.[17]

* * *

The night of September 1, 1875, Kerrigan arrived in the Lansford area with two young toughs who he had brought to do the murder of John P. Jones. He wasn't a happy little man. The deal he had made with the Pinkertons in the Tamaqua Cemetery, as he and I had understood it, was that he'd lead men in an attempt on Jones' life and they'd all get caught, presumably before any harm was done to Jones. Some time later, probably at the A.O.H. Convention on August 25, McKenna must have told him that he had presumed wrong, and that Jones would have to die. The Pinkertons and Gowen needed prosecutions resulting in death sentences, not mere jail sentences, in order to establish proper precedents for treatment of the other Molly Maguires.

Kerrigan would have objected to the change of plan, though not for any humane reasons. He simply would have doubted the Pinkertons' ability to get him off scot-free if he committed yet another murder. Also, he would have been understandably afraid that the posse who was to capture him might kill him in a fit of rage over Jones.

---

15 See for example, *MJ* September 2, 1875, September 3, 1875, September 4, 1875; September 7, 1875.
16 *New York Times* and *Philadelphia Inquirer*, April 6, 1876.
17 *Evening Chronicle*, May 1876, testimony of Patrick Duffy.

After the Convention McKenna spent two whole days in Tamaqua. His reports for the period, given to the District Attorney, don't give any compelling reason for the stay. I suspect he spent most of the time trying to convince Kerrigan that he would be set free even though Jones was killed. His August 25 report does admit that the Jones affair was heatedly discussed with Kerrigan and that McAndrew (for whom I read McKenna) told Kerrigan that he would provide the men to do the job. This would have been a threat by McKenna to take over control of the operation. This is confirmed by Linden's report to Gowen for August 27, which states that McKenna himself would go with the men to murder Jones. And that is what almost happened.

On September 2, the day after the Raven Run murders, McKenna took the train to Tamaqua, bringing along two Shenandoah men whom he had talked into doing the Jones shooting. He sequestered them in a hotel and went looking for Kerrigan. When he couldn't find the diminutive Bodymaster, he must have realized that Kerrigan had gone to do the Jones job in acceptance of the deadly new condition that had been imposed. He sent his own two men back home. Later that evening Kerrigan's friend, the proprietor of the Union House, confirmed to McKenna that Kerrigan had indeed gone to Lansford to murder Jones. McKenna could have made it to Lansford in time to warn Jones even if he had to walk – it was less than five miles – but he didn't go. He stayed in Tamaqua until after the news came in the next morning that Jones was dead. Then he returned to Shenandoah.[18]

Anyway, that's why Kerrigan wasn't happy when he arrived in Lansford with his two killers on the night of September 1. He was on risky business. Then again, he had no real choice. His best bet was to perform his part of the deal with the Pinkertons so well that they'd have no excuse for not fulfilling their part. And that is what he did

---

18 See testimony of McKenna in Comm. v. Doyle and Comm. v. Kelly reported in the newspapers, and the transcript in Comm. v. Campbell.

on September 2.

He didn't keep his shooters off the streets and away from people, as he had been careful to do in killing officer Yost. This time he processed with his men through the streets of Lansford and adjacent communities like some pagan high priest displaying a pair of sacrificial bulls. He made certain they'd be recognized.[19]

He took them to a mine where Jones worked and they talked with a mule driver who assured them that Jones was the boss there. When Jones walked out of the colliery office they were seen nearby. They engaged miners in conversation, telling one he was being underpaid for his production at five dollars per yard and should be receiving nine dollars. They were observed by witnesses near the oil shed, the lokey house, and the carpenters' shop. This was only marginally consistent with Kerrigan identifying their victim to them, which could have been done from behind cover.

They later went on a tour of most of the Welsh bars in the area, staying away from the Irish bars. Not satisfied with only those witnesses, Kerrigan took his assassins into a store and had one of them purchase cartridges while he and the other stood by. Still not satisfied, he paraded with them outside the home of a married sister of Mr. Jones who lived next door to Alex Campbell. To top even that, they went to Mr. Jones' house and one of them asked a four-year-old boy, in the presence of his mother, if Jones was home. All of these people and others who saw the shooting the following morning, about sixty in all, identified the murderers at the trials.

Kerrigan very prudently stayed away from the Jones' place the next morning. The mine boss left his house at 6:55 and walked down a path next to a pipeline that led to the Lansford train station. He had not gone far when he realized there were two men coming in back of him. He turned to see that they were strangers, and when they drew pistols from their pockets he bolted downhill. But it was no use. They

---

19 The account which follows may be confirmed in the newspaper reports of Comm. v. Doyle and Comm. vs. Kelly and in the transcript of Comm. v. Campbell.

shot him in the back and he ran into some bushes by the side of the path and collapsed. They followed and pumped more bullets into him. Then they fled to where Kerrigan was waiting to orchestrate their escape. Again, they were seen by dozens of people, some of whom would later identify them.

The capture was the most dangerous part of the operation insofar as Kerrigan was concerned. The Pinkertons' cleverness in arranging it must have gone a long way in convincing him to go through with the crime. He walked his men west to Tamaqua instead of southerly over the mountains that would have been the more direct and uninhabited route to Mt. Laffee near Pottsville, from whence they'd come. The walk to Tamaqua required about two hours.

Earlier that morning the son of one of the leaders of the Tamaqua Committee (and a member himself) had taken a train which was scheduled to put him at the Lansford Station at about the same time Jones was shot. When he confirmed that Jones was dead, the young man promptly took a special train back to Tamaqua where he spread the word about the murder and went to his office for two hours. Then a fellow told him that he had just seen Kerrigan outside of town with two strangers. So the young man and a friend went up to Cemetery Hill with a spyglass and, sure enough, across the valley they saw Kerrigan in a clearing in the woods, waiving a white cloth at them to signal his presence. His two shooters came out of the woods to join him. The young man ran to report this, and a posse went forth and efficiently surrounded the three culprits.

The posse was led by the young man's father and a family friend who was also a member of the Tamaqua Committee. Captain Linden had told both of them the truth, presumably, about Kerrigan's involvement in the plans to murder Jones. When the assassins were captured, the two from Mt. Laffee were roughly handled but Kerrigan was treated in an almost friendly fashion. The two were strip-searched but Kerrigan was asked merely to empty his vest pockets. When asked by defense counsel if he didn't examine Kerrigan's clothes or stockings or anything of the sort, the Tamaqua Committee friend answered, "No, sir, from the fact I did not think he was in it. I did not arrest him." [20]

So the truth came out in the trials, though there was little use we could manage to make of it. I'm sure Kerrigan was greatly relieved upon his capture to receive this confirmation that the Pinkertons were keeping their part of the bargain, for the time being at least.

The utter callousness of the Pinkertons and the Tamaqua Committee in willingly, nay, eagerly murdering one of their own to advance their agenda, is a horror that still robs me of sleep at night.

---

20 Comm. v. Campbell, testimony of Wallace Guss, pg. 263

CHAPTER 14

THE WIGGANS' PATCH MURDERS

Seeing a headline announcing that Jones had actually been murdered, I had the fleeting thought that something had gone tragically wrong with the Pinkertons' plan. But reading further that the three assassins had been handily captured only hours later, I realized that everything – including the death of Jones – had gone exactly as the Pinkertons had hoped. I was thunderstruck, consumed with rage, and determined to do something about it. But what could I do, was the question.

The press spewed forth a frenzy of anti-Irish and anti-Catholic bloodlust that was eagerly imbibed by the Welsh and English communities.[1] Kerrigan as well as his two young dupes, named Kelly and Doyle, were Irish. A.O.H. membership cards had been found in their pockets. That was enough proof for the newspapers to assume they'd been vindicated in their wildest imaginings about a region-wide Molly Maguire conspiracy. No one suspected that Pinkerton and Gowen were pulling the strings. I began to realize that people like Kehoe and Campbell might be in danger.

At the kitchen table with da', ma', Mary and Dutch Henry, who would be married to Mary come Shrovetide, I vented my conscience.

"I worry whether I might have saved a life if I'd done differently – either warned Jones or told everyone about McKenna."

---

1 See *Philadelphia Evening Express*, Oct. 10, 1875; *MJ*, Oct. 11, 1875 for whether Romanism was to blame, and for an example of English and Welsh reaction.

After a moment da' concluded, "Perhaps. Perhaps not. Who can say?"

"They'd have branded you as a liar or as a madman," said Dutch Henry. "That's for sure."

"Ach!" exclaimed ma' impatiently. "This talk about 'what if' is idle foolishness! It wastes time an' energy that's better spent in thinkin' about what to do next."

"I agree," said da'.

"I realize I've got to tell people about McKenna, but how do I go about it?" I asked. "If I go to the newspapers with the story they'd never print it. They'd laugh at me and say I have no proof."

"You have the telegraph that McKenna sent to Linden about our fake train wreck," offered Dutch Henry.

"No, we have only a copy in the bellboy's handwriting. They'll claim he made it up," objected da'. "It'll get the fellow fired, but it's not enough proof."

"You'd have to produce our entire family to testify as to how they fooled McKenna," said Mary. "And even so, a lot of people wouldn't believe us 'cause we're all kin."

"And some would lose their jobs," said ma'.

"We simply aren't able to assess what the consequences of a public announcement might be to anyone," da' added.

"There must be things going on which we haven't begun to imagine," I agreed.

"But it's pretty certain that if you go to the press, the family and yourself will become as much the story as the Pinkertons," said Henry.

"That's right," Mary concurred.

"First thing is to go see Jack Kehoe," concluded da'. "He needs to know, and we need to hear what he has to say."

"That's right," said Henry. "The A.O.H. might even have information which confirms your story."

"After that, you can decide what to do," said da'.

So the next day I took the train to Girardville and walked up the hill to the Hibernian House. I told Kehoe there was business that required the utmost privacy, and he ushered me into his parlor in the residence portion of the establishment. It was a very proper parlor, having an oriental rug, a harpsichord, an upholstered couch, mahogany tables, and a variety of chairs with arm rests upholstered in red velvet, in which we sat. The dark mahogany at the end of each armrest was carved in the shape of a great mastiff's head.

When I told Kehoe that for months we had known McKenna to be a detective, his eyes turned flinty. I had thought the story – our fake plot to wreck a train that caused the Pinkertons to spend three futile nights in the woods – was somewhat amusing. Kehoe was not amused. He said nothing, but he ground his teeth and his knuckles turned white as he clutched the mastiff heads. When I told him about the Pinkertons conniving to kill John P. Jones and to hang Irishmen for doing it, his face went livid and he was breathing heavily.

He didn't interrupt my story with so much as a word, but after I finished he slid off his chair onto one knee, made the sign of the cross on himself and prayed, "Jesus, Mary and Joseph, have mercy on us all."

"I hope I haven't caused any harm by not telling you this sooner," I begged forgiveness.

He sprang to his feet, and everything he had been forcing himself to restrain burst forth. "Damn! Damn! Damn it, McWilliams!" he cursed. "Damn you!"

He grabbed me by the lapels and pulled me out of my chair and demanded, "How could you not have told me this?! How could you?!"

I towered over him but, immobilized by guilt, I didn't resist and explained, "I didn't think that detectives and policemen would injure innocent people, and I believed you didn't need to be warned. I was wrong! I'm sorry!"

"Merciful God!" he exclaimed, shoved me back into my chair and shouted, "I'm the County Delegate! I'm responsible for a lot of . . . for other people . . . you . . ."

He stopped himself in mid sentence, short of insult. Bringing his hands to his head in a gesture of frustration he said, "Excuse me . . . please . . . please . . .," and he ran out of the parlor leaving the door open behind him.

I heard him moving about in other rooms of the apartment, shoving things around and, I think, pounding his fist against a wall. Finally, there came a moaning like you'd hear from a keener at an Irish wake, "Ochone!" Then there was silence.

I sat for a time. Ten or fifteen minutes, at least. I was considering getting up and leaving when all of a sudden he returned, appearing much more composed.

"I'm sorry, Mr. McWilliams," he apologized. "I'm cursed by those storms of temper, but they pass quickly. I can talk rationally now. Please forgive me."

"Of course," I agreed. "What I told you came as a shock, I realize."

He reseated himself in the chair and said, "I once told you that I'd never trusted McKenna. That's true. Anyway, I myself have nothing to hide. But there are others – some of them my own kin – who do. I warned them to stay away from McKenna; that he was underhanded and not to be trusted. But they refused to listen. They think they're much smarter than they really are. When I was kept busy here in town trying to prevent riots from breaking out, McKenna got them involved in some nasty business. Very nasty. I learned about it later. I was assuming that if I'd known McKenna was a detective, I'd have told them so.

"But thinking about it, I'm not sure I would have told them. I'd given them every other kind of warning. It should've been enough. And I wouldn't have known – any more than yourself – that McKenna was willing to kill in order to get men hanged as trophies for his agency. So my anger was triggered by assuming perfect hindsight for myself. I'm sorry."

"Thanks for understanding," I said. "But now that we know, what do we do about it? That's what I came to ask ye."

"There are some who must be told at once. Those who are most in danger," he replied. "I'll warn the ones I know about, and ye'll do the same."

"Alright," I agreed. "But if ye can avoid revealing how ye found out, I'd appreciate it."

"Not a problem," promised Kehoe. "I'll say that an informer of mine who is a conductor on the railroad told me about McKenna."

"Good," I acknowledged. "Our lads at Locust Gap are being told that there's no work for anyone who's Irish. We don't need any additional trouble."

"Well, there's one other thing," said Kehoe. "I'm going to reveal everything ye told me about McKenna to Father O'Connor of Shenandoah. He's been especially keen on blaming everything on the A.O.H. He needs an earful. An' some others of the priests might need to warn a few of their parishioners."

"Good idea," I said and asked, "But don't ye think we should try going to the newspapers?"

"Faugh!" Kehoe exclaimed. "They won't print my letters telling them the truth.[2] Most of 'em are in the pocket of the railroads. Why d'ye think the Reading keeps running that big, costly ad in the *Miners' Journal* offering a reward for the one who set fire to its Mammoth Coal Vein, when everyone knows it was a natural gas explosion? The damned ad is a daily bribe, is why. No, we won't go to the newspapers."

"Then, for a while longer McKenna will get to keep his secret, in some quarters," I said.

"Yes, an' we'll watch him closely in the meantime," said Kehoe. "It'll be interesting to see what he does when the word gets back to him that we've found him out."

*   *   *

Merely getting out the word about McKenna wasn't enough for me. I wanted to do something to help his victims, including the two

---

2 See for example, *Shenandoah Herald* of June 8, 1876, for a letter from Kehoe dated October 10, 1875.

stooges who had been hired to murder John P. Jones. They were
guilty as sin, of course. But when their cases would be called for trial,
I'd volunteer to testify as to how they'd been entrapped by the
Pinkertons through the connivance of Kerrigan. That would serve to
mitigate their guilt somewhat, I thought. It would definitely expose
the Pinkertons for what they were. In the meantime, I planned to
prepare by concentrating all my energies upon my legal studies,
especially criminal law.

But in the meantime there were distractions. On the night of
October 5 we saw another glow in the sky from our house on Tioga
Street. It broke over the eastern horizon at about 10:30 p.m. and
blossomed to its penultimate around midnight. It was the massive
Graeber & Kemper breaker in Locust Gap, set ablaze with coal oil.
One of its owners was a member of the Tamaqua Citizens'
Committee.[3] Those owners had become so enamoured with their
legal right to hire whomsoever they wished, they couldn't be made to
understand that it was unfair, anti-social, and also very unwise to
systematically exclude Irish-Catholics. This burning would
eventually be added to the long list of arsons for which the non-
existent Molly Maguires would be blamed.

A few days later I got my hands on a publication which was so
explosive that even the coal region newspapers found it too
provocative to print. It was sent anonymously to the office of the
lawyer in Ashland with whom I was studying. The moment I laid
eyes on it I knew it came from McKenna and the Pinkertons.[4] It was
captioned STRICTLY CONFIDENTIAL and was addressed to the
"Vigilance Committee(s) of the Anthracite Coal Region, and all other
good citizens who desire to preserve law and order in their midst." It
purported to give the names of all the Irish murderers in the region.

It correctly named the squad of four men from McKenna's gang
who had shot Bully Bill. (Except Mickey Doyle was misprinted as

---

3 *MJ*-Oct. 7,1875 and *MJ*-Oct. 11, 1875 under the heading DOTS.
4 See Pinkerton, pgs. 451-455.

Mickey Boyle.)

It named Hugh McGeehan as the shooter of officer Yost and named Kerrigan, and also the lad who didn't shoot, as accessories.

Tom Hurley was listed as the murderer of Gomer James.

Kerrigan and his two dupes, all of them securely lodged in the Carbon County jail, were noted as the trio who had shot John P. Jones.

Most surprising of all was the naming of three members of the O'Donnell clan of Wiggans' Patch – two of them Kehoe's brothers-in-law – as the Raven Run murderers. (Two of McKenna's cronies were named, as well.)

I remembered Kehoe in his parlor telling me, after his temper had calmed, that some of his own kin had been lured by McKenna into some very nasty business. I realized that the "nastiness" must have been the murder of the mine boss at Raven Run. The "kin" to which Kehoe had referred were none other than his brothers-in-law "Friday" and Charley O'Donnell, named in the list of murderers. I felt certain it was accurate, and I understood fully, for the first time, why Kehoe had become so upset.

Assuming that Kehoe had not yet seen the STRICTLY CONFIDENTIAL memo, I immediately made a copy of it and rented a carriage and drove to Girardville to warn him about it. But he wasn't home. His wife told me he had gone to Pottsville. That was too far for me to follow him in my buggy. But Wiggans' Patch was only six miles east. I drove there to warn the O'Donnell boys myself. I wouldn't be tardy this time.

The house of the widow O'Donnell, the mother of Kehoe's wife and matriarch of the clan, was one-half of a large duplex. The neighbors told me it housed her daughter's family, the brother of her daughter's husband, her unmarried sons, Charley and Friday, herself and three unrelated men who were paying bed and boarders. However, when I got there only one person seemed to be at home – a shifty-eyed, unshaven lad of about nineteen years who opened the back door to the kitchen after I rapped on it persistently.

"Who are ye? What d'ye want?" he grumbled and glared at me suspiciously. He wore only trousers and an undershirt.

"My name is Matt McWilliams. Used to work for the W.B.A. I need to talk to Friday and Charley O'Donnell," I said.

"I've heard of ye. Come in," he stepped out of the doorway and motioned for me to sit at the kitchen table and admitted, "I'm Charley."

He also sat, and he spooned himself a big mouthful of porridge from a half-full bowl and mumbled, "Say yer peace. I'm listenin'."

"I came to warn ye," I said. "D'ye know that Jim McKenna, the head man up in Shenandoah, is really a Pinkerton detective?"

"Yep, we heard about that some time back, from a conductor on the Reading," he said through another mouthful of gruel.

"No, ye heard it from your brother-in-law, Jack Kehoe," I corrected him. "I was just in Girardville to see him. He wasn't home, so I came here to warn ye myself."

"About what?" he asked, his interest aroused for the first time.

"About the Raven Run murders. Someone, probably the Pinkertons, is spreading the word that you and your brother Friday and your sister's husband all had a hand in it. Along with two of McKenna's cronies."

"Damn! Is that so?!" he exclaimed. "We didn't figure on him doing that."

"Why not?" I asked. "Don't you realize he fully intends to have you all arrested, convicted and hanged?"

"You're wrong. They can't put us on trial," he said with confidence and filled his mouth with the last of the mush.

"Why not?! Why in heaven's name can't they?" I demanded.

He stared at me a moment, then flashed me a big, smug smile through his gapped teeth and answered.

"The minute we learned that McKenna was a detective, we went to see him. All three of us. Our talk was very private; out of hearing of even that weird bodyguard of his. We told him we knew who he was. An' if he dared to have us arrested for the Raven Run job, then

all three of us would get on the witness stand and admit it, since we'd
hang for it regardless. But then we'd also testify that he was the leader
of our group an' the one who took the first shot! We'd make certain
that he'd hang right along with us!"

"McKenna was with you?!" I exclaimed.

"Yeah. It was the three of us along with him an' his buddy, Tom
Hurley."

"My God!" I exclaimed as the weight of McKenna's murderous
villainy was dropped on my conscience yet again.

"Here! Read this!" I groaned, handing my copy of the STRICTLY
CONFIDENTIAL handbill to Charley.

It took him a while, but when he finished he had mastered its
meaning.

"Christ!" he mumbled, handing it back to me. "Almost all o' the
men named here are either in jail, or they've fled the coal region or
they're McKenna's own gang. Assumin' they're all safe, the only ones
left for the Vigilantes to get are us O'Donnells."

"That's the way I see it," I told him. "That's why I came to warn ye
directly."

"Thanks," he said. "Can I fetch ye a drink?"

"No," I declined. "I don't want ye to think that I approve of any
part of what ye've done. I don't, at all."

"Why not?!" he exclaimed defiantly. "That son-o-bitchin' boss
was worse than any Orangeman grown in Ireland. He deserved it."

"No," I disagreed. I rose from the table but paused on my way out
the kitchen door to add, "No one deserves to be shot down like that.
Not even the likes o' you! That's why I warned ye!"

I slammed the door shut behind me.

*   *   *

From what I understand, the O'Donnells were quite careful of
themselves over the next two months. They didn't go anywhere
unless they went in a group, and then they were sure to be heavily
armed. After a while, though, Charley became cocky and is reported

to have bragged that none of the Modocs had the moxie to face up to him and his clan in a gunfight.

The STRICTLY CONFIDENTIAL handbill was widely circulated among the "respectable" citizens of the County and, for the most part, they accepted it as gospel truth.

The *Shenandoah Daily Herald* persistently published broadsides that demanded that Lynch Law be used to root out the Molly Maguires whose identities, it hinted, were known. It claimed that such justice would have been meted out already if not for the milk and water policy of the *Miners' Journal* and some other papers. The *Journal* flatly denied that it would ever say "hold" to anyone putting a rope around the necks of the Raven Run murderers, although it would prefer to see the sheriff adjust the noose.[5] Kehoe wrote letters denying that the A.O.H. was involved in any way and pointing out that the Mollies, to the extent they existed, were merely unassociated gangs of tavern toughs.[6]

At about three in the morning of December 10, 1875, a Vigilante Committee, probably organized by the Pinkertons, struck.[7] There were at least thirty of them. If some were Coal & Iron Police, their uniforms were covered by rain slickers that extended below the knees. The Vigilantes, for the most part, were local men who could identify the O'Donnells. They wore masks over their faces.

They knew the layout of the widow's four-room duplex, the first floor having a front bedroom and a kitchen in the rear. A doorless stairway went from the kitchen up to a large front bedroom, which was passed through to reach a smaller back bedroom where the widow slept alone.

Without warning, the Vigilantes smashed down the kitchen door and rushed upstairs to the large room where the three unmarried men of the clan shared their beds with three paying boarders. They caught everyone groggy with sleep, and pointing pistols in the men's

---

5 *MJ*, Oct. 24, 1875.
6 See *Shenandoah Herald*, June 8, 1876; *MJ*, Oct. 24, 1875.
7 *MJ*, Dec. 11,13,14 and 15, 1875. See: Letter of A. Pinkerton to G. Bangs, Aug.29, 1975.

faces they began to sort out who was who.

The noise woke up the widow's pregnant young daughter and her husband in the downstairs bedroom, which had a door leading to the basement. The husband knew he had been mistakenly named as a Raven Run murderer, confused for his brother sleeping upstairs. He whispered to his wife, Ellen, to lie still and he scurried down into the basement to hide. Ellen didn't obey her man. In her white nightgown she hushed her three-year-old son, then got out of bed and opened the door to the kitchen to learn what was happening. Seeing her form appear in the doorway, one of the Vigilantes fired, his bullet piercing her breast.

"I'm shot!" she cried and fell to a heap on the bedroom floor. Her baby screamed and hid his head under the covers.

Upstairs, the widow O'Donnell also had heard the noise of the door breaking. She had the notion that one of the boarders, a feeble old man, might have fallen down the stairs to the kitchen. She got up and opened the door between the bedrooms to find the front one packed with armed men. They had a lantern.

One of them, grown impatient with his mask, had taken it off and dropped it. She recognized him as a butcher from Mahanoy City! She had done business with him for almost ten years! When he saw her looking at him, he pointed his pistol at her, but another Vigilante grabbed his arm and said, "No!"

The butcher didn't shoot, but he hit the widow on the side of her head with the butt of his pistol. She reeled back into her room and fell, stunned, upon the bed.

The Vigilantes quickly identified the palsied old man and a younger man as boarders. They left the old fellow alone but tied the younger to the bed. The four other prisoners they took downstairs and out into the street in front of the house. There, they asked for names, and when satisfied they had identified the third boarder, they ordered him to return to the house. He went willingly, but one of the Vigilantes fired at him, the bullet passing through his nightgown but missing him.

The unexpected shot startled the other Vigilantes and the three O'Donnell clansmen saw the opportunity to break free. Two ran away to either side of the highway. Shots were fired at them and one took a bullet in the arm, but they both escaped. Charley O'Donnell had opted to flee directly down the road. A bullet struck him and he rolled onto his back, moaning with pain.

Almost as though they were enacting a ritual, the Vigilantes walked over to him without hurry and each delivered a single shot into his body. He absorbed more then twenty bullets from his groin to his head, some delivered so close that they set his nightclothes afire.

The Vigilantes then walked off in various directions and were enveloped by the night. A neighbor woman ran out and ripped the burning clothes off the corpse.

CHAPTER 15

THE "SHOWTRIALS" BEGIN

The coroner came from Mahanoy City and examined the bodies of Charley and his sister. He brought with him some men whom he appointed as a Coroner's Jury and he commenced interrogating witnesses right in the widow O'Donnell's front bedroom. He was asking the widow whether she could identify any of her attackers when Jack Kehoe stormed in. Not sure of what was going on, suspicious of the Coroner and his personally selected Jury, and fearful of putting a noose around the necks of his in-laws, he ordered the widow, "Don't tell 'em anything! This business will be settled in another way!"

The Coroner ordered Kehoe to be quiet or to be arrested, but the widow obeyed her son-in-law. The Coroner persisted to ask her whether she knew any of the attackers. She said she could not say. Then she said she didn't know.

Later that day, with Kehoe's approval, she brought charges against the butcher whom she'd recognized as one of the Vigilantes and the man was arrested. But he was treated as a hero by the non-Irish citizens of Mahanoy City. It was said that if he'd needed a bail of a million dollars, they'd have raised it for him. Even prior to his preliminary hearing, the *Miners' Journal* announced that his friends were certain he'd be discharged, "So the only result from his arrest will be the mortification of passing two days in prison."[1]

---

1 *MJ*-Dec. 13, 1875

The *Journal* was correct. The widow's testimony had been tainted by her statements to the Coroner, the Judge ruled. The butcher was set free amid the jubilation of a great throng of people. Most of them were celebrating not because they thought the fellow innocent, but because they heartily approved of what he had done.

The two men of the O'Donnell clan who had escaped the Vigilantes were reported to have fled to Ireland.[2] They were never again seen in the anthracite region. Thus, McKenna and the Pinkertons were rid of all witnesses who would have implicated them in the Raven Run murders.

On December 21, 1875, the *Miners' Journal* published the final piece of Gowen's fabrication against the coal region Irish. It came from his friend and confidant, Archbishop Wood, an Englishman converted to Catholicism. The cleric didn't exactly dislike all of his Irish flock, but he certainly mistrusted them, and their centuries old refusal to embrace British domination. Wood issued a pastoral which excommunicated "the Molly Maguires, otherwise the Ancient Order Of Hibernians." Thus, Mr. Gowen's studied ambiguity about who and what were the Molly Maguires – an ambiguity vital to his successful prosecutions – was imprimatured by the very Church of his victims.

The stage was set for the showtrials of the Molly Maguires, which would dominate the nation's news coverage for the next two years and mark the pinnacle of Gowen's power over business and life and death in the anthracite region.

*   *   *

Shortly after New Year, I went to offer my testimony to the attorney heading up the defense team in the first of the Molly trials, to begin January 18. He was Lin Bartholomew, Esq., a veteran of seventy-seven homicide trials. I wished that it had been someone else. The fellow didn't strike me as being very bright. He'd been a prosecuting attorney against Dan Dougherty and I didn't believe he

---

2 *Pottsville Evening Chronicle*, Feb. 12, 1876.

had really understood that case. Also, unfortunately, he had accused me of telling Dan's witnesses what to say, and uncordial words had passed between us.

As I unfolded to him my story about McKenna being a Pinkerton detective who had forced Kerrigan to hire his client to kill John P. Jones, he drummed his fingers on his desk impatiently. Finally, he interrupted.

"So far, McWilliams, everything you're saying makes my client guilty!" he growled. "Do you have anything that's exculpatory?"

"I can't prove yer client's innocent," I told him. "From what I hear, no one can. But I can throw a light on the scene that makes him less guilty, and places the heaviest blame on Gowen and his Pinkertons, where it really belongs. The truth has got to be told!"

"You've yet to learn, boy, that a trial isn't about the truth," he smiled snidely as though instructing a simpleton. "It's about guilt or innocence. My client's plea is 'innocent'. I can't argue that he was talked into doing the deed, not even if by the devil. It's no excuse!"

"But there's the defense of entrapment!" I protested.

"That's a defense only if officers of the State caused him to commit the crime," he explained irritably. "The Pinkertons are a private agency."

"All right, but Captain Linden is a member of the Coal & Iron Police," I argued.

"They also are a private force," countered Mr. Bartholomew. "Read your statute, boy. If you testified, the Judge would throw out your defense and use your words to hang my client!"

"But, sir," I argued, "in this region where all the land, even the minerals under the streets and houses and public buildings, belong to the Railroad or the coal companies, a private force policing that property is *de facto* exercising the police power of the County."

"Judge Dreher would laugh at your Latin and dismiss your argument out of hand," the lawyer chided.

"Sir!" I pleaded. "What ye say may well be so . . . or rather, is probably so. But it shouldn't be. At least, we'd have a chance to reverse the Judge before the Court of Appeals. Regardless, all I ask is that ye think about whether what I told ye might in some way be useful to these men."

He leaned back in his chair, regarded me with distaste and quipped, "What would be helpful is if you would stop wasting my time."

I walked out of there with my face and ears turning red. I felt shamed that Bartholomew had scoffed at me. I felt angry, because I believed that even a gob of a lawyer like him should have been able to figure out some way to use the information I was offering.

But it was not to be. I didn't personally count the prosecution's witnesses, but the papers said there were a hundred-twenty of them and I believe it. No motive for the shooting of Jones by Mr. Bartholomew's client was proved, but it wasn't necessary. Kerrigan had made certain that the hapless defendant could be identified conclusively, and the murder weapons had been found where he'd been skulking in the woods. His lawyers offered no testimony – no evidence of any kind. Anyone Irish had been excluded from the jury panel and public excitement against the defendant had been whipped up by the papers to fever pitch.[3] The outcome was foregone, and on February 1 the verdict of guilty was duly entered.

The newspapers were delirious with joy that the first of the Molly Maguires had been convicted. This was in spite of the fact that the name Molly Maquire had never been uttered in the courtroom, to the best of my recollection.

On February 4 Alex Campbell was arrested, charged with helping Kerrigan to murder John P. Jones. On the same day four other men – two alleged assassins and two alleged planners – were arrested for helping Kerrigan to murder officer Yost. Several days later one of the O'Donnell clan and one of McKenna's gang were arrested for the

---

3 Editor's Note: See *The Hard Coal Docket*, by John P. Lavelle for a thorough and compelling study of the biased jury selection process in Carbon County in 1876 and of the undue excitement and prejudice against the Carbon County defendants which prevented them from receiving a fair trial.

Raven Run murders. The *Shenandoah Herald* gloated, "More Mollies Scrapped Up" and the *Miners' Journal* gleefully parroted, "More Mollies Gobbled."[4] The existence of an organization named the Molly Maguires was taken for granted.

Also, the newspapers shamelessly prejudged the defendants. The *Pottsville Evening Chronicle* of February 5 stated, "That the officers have secured the right men, there appears to be little doubt . . . The detectives . . . have gained . . . important evidence which is said to be strong enough to convict the whole gang . . ." It was becoming impossible for Irish defendants to obtain an impartial jury in the coal region.

On February 12, a preliminary hearing was held to see if there was sufficient evidence to try the men accused of murdering officer Yost. In open court, Gowen's lawyers announced – for they had taken over the prosecution from the Commonwealth – that Kerrigan was an informer. However, it was falsely indicated that Kerrigan had joined forces with Gowen only after he was put in jail for the murder of John P. Jones. A confession signed in jail by Kerrigan was read into evidence. The *Evening Chronicle* of February 12 reports Kerrigan's first words of confession as follows: "Campbell handed me a pistol and wanted me to go along and kill Jones." In fact the whole confession had nothing much to say about the death of officer Yost. Its sole purpose seemed to be the identification of Alex Campbell as the chief of those who conspired to murder mine boss Jones. It seemed strangely out of place in the Yost case, wherein Campbell was not a defendant.

At about this time the word got back to McKenna that within the A.O.H., it was becoming widely known that he was a detective. He at once went to Kehoe and tried to bluff and fluster the truth away. He demanded a special meeting to hear his defense. But Kehoe sent him to see Father O'Connor to obtain a summary of what was known about him. The priest told him that not only was he known to be a

---

4 *Shenandoah Herald*, Feb. 10, 1876; MJ-Feb. 11,1876.

detective, but he was known to be a dishonest one. He was cognizant of crime long before the perpetration of it. He was, in fact, a stool pigeon who knew all about crimes and took part in them instead of reporting them and preventing them as a detective should. Faced with this incontrovertible conclusion, McKenna fled the coal region that very day, March 7.[5]

When I read about the confession of Kerrigan, I suspected what Gowen was up to. I swallowed my pride and went again to Attorney Lin Bartholomew because again, unfortunately, he was lead defense counsel for the second of Kerrigan's dupes to be tried for the murder of John P. Jones.

"You've got to let me testify," I pleaded. "This time ye'll not only be facing the same one-hundred-twenty witnesses as in the first case. Ye'll have the testimony of Kerrigan to contend with that yer man shot Jones. McKenna might also be brought in to testify. Ye have nothing to lose by letting me tell the truth about . . . ."

"Why isn't that a good thing?!" the lawyer interrupted me.

"What's a good thing?!" I asked.

"Kerrigan in the case," he answered. "The fellow's a weasel. He even looks like a weasel. If I can make the jury see that he's a vicious thug, they might have some sympathy for my boy."

"Yer boy is less than twenty-years-old, an' already he's earned himself a bad reputation," I replied. "The jury won't be sympathetic. But they'll be shocked if ye let me testify that it was Gowen and the Pinkertons – the very prosecutors in this case – who hired him to commit the murder. Maybe they'd be outraged enough to set him free."

"No, no! No, no!" the lawyer exclaimed. "I won't adopt a defense which requires an admission of guilt. I won't do that. The Judge would advise the jury that the defense is invalid, and instruct them that they must find the defendant guilty. The jury wouldn't disobey such an instruction no matter how outraged they might be."

---

5 Comm. v. Kehoe,et.al.,testimony of James McParlan,pg.97.

"It's yer only chance!" I warned him. "Anyway, Kerrigan isn't going to take the stand to testify about yer boy. The prosecution has their one-hundred-and-twenty witnesses who will do that. Read Kerrigan's confession. The real reason for his testimony will be to claim that Alex Campbell ordered the murder."

"So what?! That could help this defendant," he countered.

"No it won't," I argued. "Campbell's neck is on the line here, as well as yer client's. His trial will come up soon. But if ye don't let me reveal that it was the Pinkertons who controlled Kerrigan, then Campbell will be convicted in the press before his jury is picked."

And so it was. The lawyer wouldn't present my testimony. Because Campbell wasn't a named defendant in the trial, Kerrigan was allowed to testify without serious challenge that it was Campbell who ordered and assisted the defendant to murder Mr. Jones. Thereafter, the newspapers took it as "common knowledge" that Campbell was the primary planner of the assault. He was doomed, I feared. And that proved to be no help to the young man who was the defendant. He was found guilty of murder in the first degree.

*   *   *

Next to be tried were the men implicated by Kerrigan in the murder of officer Yost. Only Kerrigan's testimony was absolutely necessary for their conviction, for he was an eyewitness and his story was corroborated by a miner who had encountered the two assassins on their return home from the kill. Nevertheless, in the prosecutor's opening statement, it was publicly admitted that McKenna was a detective. He would take the stand and testify that all of Kerrigan's alleged abettors had confessed to him their role in the murder of Yost. Thus, the detective who allegedly was never to have testified would now testify, even though it wasn't necessary.

The reason why McKenna was being called to the stand was made clear. The District Attorney announced that the prosecution would disclose, through McKenna, the terrible doings of the banded assassins known as the Molly Maguires. And the identities of the assassins were dramatically revealed when officers came into the courtroom and spread the news that Jack Kehoe, and all of his A.O.H. Bodymasters who had attended the June 1 meeting to plan the miners' strike march, had been arrested for planning to shoot Bully Bill Thomas! The Mollies were labor agitators!

The *Evening Chronicle* understood these startling events in the courtroom and offered this explanation:

> *"What does it mean?" was an oft-repeated question and it seemed that everybody that heard the news realized the fact that the day of the DOOM OF MOLLYISM was here . . . a more important event than the trial in the Court.*[6]

Gowen himself would preside as chief prosecuting attorney. He wanted to prove that the Molly Maguires actually existed, as most people already supposed, and that anyone who was a leader of the Mollies, otherwise the A.O.H., was probably guilty of complicity in any murder for which he might be charged. For this result, McKenna's testimony would be absolutely necessary, as had been obvious from the beginning of his employment.

The Court had publicly advised the jury panel that it was their duty to help the District Attorney purge the terrible murders that had disgraced their county.[7] The Court could do no less, and it overruled the defense's most proper objection that testimony about Molly Maguires wasn't relevant for understanding Kerrigan's simple plan to seek revenge upon a police officer.

For four days Mr. Gowen was unleashed to direct McKenna's testimony; how he had come to the coal region and learned about the Molly Maguires, otherwise the A.O.H., otherwise the Buckshots, otherwise the Sleepers, and otherwise whatever Gowen wanted to call them. McKenna wisely admitted that the stated purposes of the A.O.H. were benevolent, and that the rank and file didn't know about the secret cabal of Bodymasters and their henchmen who were the

---

6 *Pottsville Evening Chronicle,* May 6, 1876.
7 MJ-Sept. 7,1875.

true Mollies. Still, he continued to disclose the A.O.H. passwords, organization, signs, practices and meetings as though they were Molly Maguire devices. He was permitted to infer, without any proof, that the entire A.O.H. was tainted. Defense counsel didn't cross-examine him about these inconsistencies because the defendants weren't accused of being Mollies, but of being murderers. The upshot of it was that Gowen fulfilled his boast – that he would make the A.O.H. so odious that to be a member of it would be *prima facie* evidence of guilt.

Of course, McKenna did testify as to the issue being tried, the murder of officer Yost. He said that one of the shooters and all the planners, including Bodymaster Roarity, had confessed to him that Yost was murdered in return for Kerrigan's promise to kill mine boss Jones. The problem with this simple story, from Gowen's perspective, was that Alex Campbell was a completely unnecessary party to the alleged conspiracy. The conspiracy already contained defendant Roarity as a planning Bodymaster from Campbell's locality. Campbell was a fifth wheel.

Indeed, on the first day of McKenna's testimony, I thought he might be exonerating Campbell. He related that on August 4, 1875, he and Kerrigan stopped at the Union House in Tamaqua. Campbell was there doing his wholesale liquor business with the proprietor. It was only after Campbell had left in his wagon that the proprietor and Kerrigan gave him the names of McGeehan and his buddy as the ones who had shot Yost.[8] I breathed a sigh of relief. If McKenna had wanted to implicate Campbell in the plot, he would not have admitted that the plotters had waited for Campbell to leave before naming their accomplices.

I was wrong. Three days later McKenna completely changed his story. He testified that in the Union House on August 4, it was Alex Campbell who told him that McGeehan was the one who shot Yost. Moreover, Campbell then said he would reward McGeehan by

---

8 *MJ*-May 6, 1876, test. of McParlan.

starting him in a saloon, and also Campbell said, "He wanted a few good men to go over and shoot John P. Jones." Finally, Campbell related that he had prevented Kerrigan from going over on the 27th of July to shoot Jones because Kerrigan, being a very small man, was too easily marked.

The mind-numbing contradictions between the two stories need no comment. However, it won't belabor the obvious to note that McKenna had already placed the phrase about wanting "a few good men to go over to . . . murder Jones," into Kerrigan's mouth.[9] Also, if Campbell had stopped Kerrigan on July 27 from going to murder Jones because the little man was too easily marked, he wouldn't have harbored Kerrigan the day before the murder and allowed him to traipse all over town.

What's even more mind-numbing is that McKenna seldom told a consistent story about an event, whether in a single or different trials. Yet, he consistently got away with it. In this first Yost case the defense lawyers, led by my nemesis, Lin Bartholomew, refused to cross-examine him about these discrepancies. They said they weren't relevant to the issues at hand, because Campbell wasn't on trial, and cross-examination of McKenna about them would only afford him additional opportunity to repeat his testimony against their clients.

The only light shown upon McKenna's machinations came when one of the defense counsel cross-examined McKenna about his failure to warn John P. Jones, Bully Bill Thomas, and the Raven Run mine boss about the assaults which had been planned upon them, and thus to prevent the crimes. McKenna had not possessed pre-knowledge about the murder of officer Yost, but such cross-examination was relevant to McKenna's credibility as a detective. McKenna said he didn't warn the three men because he was afraid of losing his life. This was a lie but the jury and, most importantly, the press believed it. They might not have believed it if the defense had challenged the detective's other demonstrable lies, even if they were

9 *MJ*-May 9, 1876.

about Campbell. To this day I remain stunned by the flagrant incompetence of the defense lawyers.

I now know after reading McKenna's expense reports that his testimony in the Yost case was full of other lies about Campbell. He swore that on July 15 he went to Campbell's saloon to find out who had shot Yost. His expense report shows he had not gone to Campbell's; rather he had actually gone to a different saloon. He further swore that on July 17, he went to Campbell's and got information from his son. The expense account shows he actually went to a different saloon and got information from that saloonkeeper's son. He testified that after the party to celebrate the opening of McGeehan's saloon, he went back to Campbell's place. The expense report shows he went to a different bar entirely.[10]

Thus did McKenna lie throughout the Yost case for the purpose of convicting Alex Campbell and the nebulous Molly Maguires in the court of public opinion. Not only the coal region newspapers, but also some of the big city papers, printed a summary of the testimony of McKenna in its entirety, along with comments about how conclusively it proved the prosecution's case. To make matters worse, one of the jurors died. The case was retried with a new jury. The reading public was again bombarded with McKenna's lies and the rhetoric of Gowen's lawyers and encouraged by the press to accept them as true.

Between trials, the newspapers continued to wax even wilder and more frantic about the Molly Maguires. An assault involving some Irishmen in far away Blair County was reported in the May 25 *Pottsville Evening Chronicle* with the following conclusion:

> There seems to be no longer any doubt of the existence of a well organized gang of Mollies, or murderers, at Gallitain.

It might have reported the bloody Custer Massacre of that June as a depredation by the Molly Maguires had a single Hibernian been

---

10 See *MJ*-May 6,1876 and May 9, 1876 and compare with McKenna's expense reports for those days.

spotted among the Indians. In any event, the Molly Maguires, whoever they might be, were indeed doomed in the hard coal region. The Court's improper admission of irrelevant evidence and the demagoguery of the press about that evidence had sealed the fate of anyone accused of being a Molly.

CHAPTER 16

ALEX CAMPBELL AND HIS JURY

It was a rainy morning in early June when I approached the Carbon County Jail to meet Alexander Campbell for the first time. The massive stone keep had been built only five years previous, but already it was nicknamed the "Bastille." Two tall cellblocks and an even taller tower loomed over the wet street behind a twenty-foot stone wall surrounding the place.

I had a sense of foreboding as I splashed up the flight of stone stairs giving access to the double-doored entrance. The four newspapers of Carbon County had managed to make Campbell the most reviled man in the region by publishing Kerrigan's confessions and by attesting to their total acceptance of the word of Tamaqua's most notorious villain.

The little Bodymaster's statements had negated his own role in the Yost-Jones murders and blamed everything on Campbell. Kerrigan claimed that he and the gunmen had gone first to Campbell's Hotel on the night of September 1 to seek instructions about murdering Jones, and then all four had walked to McGeehan's saloon to inspect the pistols to be used. The next day he took the men back to Campbell's for supper and afterward the two shooters went out alone looking for Jones but couldn't find him. They returned and slept the night at Campbell's. Finally, he said, he and the gunmen left Campbell's the next morning with orders from Campbell that they must kill Jones as the boss left his house to go to work.

If true, this account was devastating to Campbell. It had been leaked to the press by the prosecuting attorneys. In gratitude, the newspapers made clear that Kerrigan was completely believable, that "Alex Campbell was the greatest Roman (i.e.-Irish-Catholic) of them all", and that "MURDER WILL OUT."[1]

I opened the front door to the jail, entered a foyer, and advised the guards that I had arranged to visit Mr. Campbell.

His lawyers had refused to meet with me. They'd been warned by Attorney Bartholomew that I was no more than a courthouse gadfly. But Mrs. Kehoe had gotten word to Mrs. Campbell that her husband should see me, and so he'd agreed.

The guards escorted me into a cellblock. It was starkly and brutally simple – a wide corridor two stories high with five solid cell doors on each side of the first story and five more on each side of the second story. An iron staircase rose to an iron balcony wrapping around the room at the upper floor level to give access to the ten cells on the second story. The stone walls were windowless and the doors were of thick wood.

The door to Campbell's cell, number seventeen, was opened for me. It was small – claustrophobic – with a single pitiful window less than six inches wide and about thirty inches high.

Campbell had not yet been put in chains. He was a big, broad man and handsome in my opinion. After I introduced myself, he sat on his cot and I perched upon a stool.

"I've come to offer ye my testimony," I told him.

"I know," he replied. "My attorney says ye'll testify that McKenna and Captain Linden offered Kerrigan his freedom if he'd bring men to shoot Jones and if he'd make certain they'd be caught."

"That's right," I acknowledged. "I believe that revealing their unholy bargain would be very useful to yer defense."

"My attorneys don't think so," he said.

"Why not?" I asked. "I don't understand."

Campbell explained, "My lawyers can prove that on the night before Jones was murdered McKenna came to Tamaqua looking for Kerrigan and bringing two of his own men who were willing to help the little squirt do the job. So without your testimony we can prove that McKenna was encouraging Kerrigan. We don't need it."

"That's good! That's very good!" I exclaimed. "But that doesn't prove that McKenna was encouraging Kerrigan as an officer – a detective – with an offer of freedom."

Campbell peered at me with raised eyebrow, as though impressed with my understanding of the matter, and he said, "That's true, to be sure. But my lawyers would rather not create the impression that Kerrigan and the Pinkertons had made such a deal."

"Again, why not?" I demanded.

He sighed and sat looking at me for a long time, as though making a decision about me. Finally he asked, "D'ye believe I'm innocent?"

"I'll believe it if ye tell me ye are!" I shot the answer back to him.

He smiled, albeit a little sadly, and replied. "Ye're a smart lad. I trust ye. Yes, I'm innocent. But not entirely innocent."

"What d'ye mean?!" I demanded.

"I knew nothing about the Yost murder before it happened," he answered. "Afterward that piss-ant Kerrigan went around bragging about it and of course I found out. A lot of people knew. I even discovered that some of our local lads had been involved. I suspected there might be more trouble to come but I didn't want to know any more, and I tried to avoid knowing. Kerrigan and I never were on very good terms. He wouldn't have dared to bring his gunmen into my house, because he knew I would have realized what they had come for, and he knew I would have stopped them. And I would have, too! But they never came into my place. Never!"

"So ye are innocent!" I exclaimed.

"Yes, except I'm guilty of knowing too much and not doing anything about it," he qualified. "We Irish are raised to detest informers, don't ye know. I feel guilty about Jones, even though he was a blacklisting bastard."

"Then Kerrigan's claim that ye fed an' bed him an' his men is totally false?!" I declared.

"Totally," he confirmed. "An' what's more, we can prove it. Conclusively. We've found a man who was in McGeehan's saloon all evening on September l, an' he'll attest that I never came in; nor did any of Kerrigan an' his men. Three of my own customers will swear that Kerrigan and his men didn't come into my place that evening. My cleaning maid and a boarder will say the same. Six witnesses will swear that the murderers were never under my roof on September 2, including a boarder who had supper with us who will verify that the killers weren't at the supper table that night as Kerrigan claims. There's even a Justice of the Peace from Tamaqua who will prove that Kerrigan lied when he reported meeting the Justice on his way to my place. Finally, thank God, there was a man in my hotel early on the morning of the murder. He was seen here by both of Mr. Jones' sisters who live next door. He'll confirm that the killers didn't set out from my place as Kerrigan claims. Hell! If they had, they'd have been seen by both of Jones' sisters, who saw my witness!"

"This is wonderful news!" I declared. "But are they truly independent witnesses?"

He shrugged and replied, "The hotel is my place of business. Most of them are my customers, boarders, employees or my in-laws. They aren't A.O.H. members, if that's what ye mean."

"Good; that's good," I said. "But that doesn't explain why your lawyers don't want me to tell how Kerrigan had made the deal with the Pinkertons."

"They're worried that the jury might take it wrong," he answered. "If Kerrigan was working on behalf of the mine operators then, definitely, he'd have wanted to involve me by coming into my hotel.

The jury might conclude that Kerrigan actually did what his employers must have been prompting him to do – and my witnesses are lying!"

I was stunned by the argument – Kerrigan must have brought the men into Campbell's because his bosses would have demanded it! That theory had never entered my mind.

"Well, that is certainly one possible inference," I submitted. "I'll have to think about it. We wouldn't want that."

"Also," Campbell added, "my lawyers say that we already have a good, strong defense. An' when that's so, it's best not to complicate it by trying to prove more than is necessary."

"That's true; that's true," I had to admit aloud. But I had the uneasy feeling that, once again, truth was not being treated as a pole star and that without it, Campbell's defense might founder.

*　　*　　*

Two weeks later, the trial commenced.[2] Hoping that Campbell's lawyers would change their minds about using my testimony, I attended every session. Mrs. Campbell found that I was useful to explain to her various points of law as they arose, and she invited me to sit next to her in the front row of spectators.

She was an unadorned but beautiful woman in her mid-thirties. Her intelligent face, noble restraint, and graceful bearing conveyed an aura of rectitude which mirrored that of her husband.

On the beginning of the first day the panel from which the trial jury would be selected poured into the courtroom. About fifty men were seated in pews usually occupied by the spectators during trials. Each was called to the witness stand to answer questions in the presence of all the others. The opinions and prejudices of each would be revealed in front of all.

I noted their clothes, faces and demeanor and listened to them

---

2 Editor's Note: For a penetrating analysis of the trial testimony and an intriguing investigation into the life of McKenna see: *A Molly Maguire Story* by Patrick Campbell, grand-nephew of the defendant.

speaking.

"My Go---goodness!" I exclaimed to Mrs. Campbell. "There's not a single Irishman! Have we all been excluded?!"

"I've never seen any of these men before," she replied. "Quite a few were talking German."

The second one questioned admitted that he had read Kerrigan's confessions. He said the newspapers had convinced him that Campbell was guilty and, "It would take strong evidence to change my opinion about the defendant's guilt."

The defense challenged the fellow for cause and the objection was sustained and the fellow removed.

The next juror claimed that he had read nothing in the newspapers, and he was seated.

The fourth man admitted that he'd read Kerrigan's confessions as well as all the testimony in the other trials. He said he had formed a strong opinion about Campbell's guilt and it would take strong evidence to change his opinion. At this, Judge Dreyher interjected and asked him to state whether he could render a verdict uninfluenced by opinion.

He answered, "I think I could."

The defense challenged the fellow for cause. The Court overruled the objection and he was seated in the jury box.

"This is terrible!" I groaned to myself.

And so it was. Seven of the twelve jurors had read "Kerrigan's Confession," so damning to Campbell. Eleven were thoroughly familiar with Campbell's alleged participation in the exchange of the Yost murder for the Jones murder from reading McKenna's and Kerrigan's testimony in the Yost case. Four of them admitted that they believed Campbell was guilty. Equally prejudicial was that all the jurors in the box were exposed to the dismissed members of the panel openly proclaiming the opinion of the non-Irish communities that Alex Campbell was guilty.

A manager of the biggest coal company in the County advised his brethren, many of whom were prospective employees, that Campbell was guilty and, "There ought to be more hangings than there is!"

The publisher of the virulently anti-Irish *Mauch Chunk Democrat* boasted that he had written editorials about Campbell's guilt and, " … of course Campbell is guilty!"

Twelve other panel men were permitted, in similar fashion, to pressure their peers into finding Campbell guilty.[3]

I looked into the grim faces of the jury on the first day of testimony and I prayed that one or two of them would be able to discern the truth in spite of the inflamed passions of our times.

Mrs. Campbell clutched at my arm and whispered despairingly, "Their minds are already made up!"

"But our defense is strong," I reassured her. "If Mr. Gowen had believed that conviction was certain, he'd be here to try the case himself."

I was expressing more optimism that I actually felt, trying to help her through her ordeal. She recognized this, smiled her appreciation, and said, "Thanks."

---

3 The testimony quoted in this chapter and the next is from the transcript of Comm. V. Campbell, for the murder of John P. Jones – Carbon County. Editor's Note: Said transcript had been lost and was rediscovered in the early 1980's in the basement of the courthouse.

CHAPTER 17

TRIAL OF ALEX CAMPBELL

The prosecution put on a multitude, over six dozen witnesses, who recounted the antics of Kerrigan and his henchmen the day before the murder of Jones and on the fatal day. Then Kerrigan took the stand. He seemed to have a high opinion of himself as an important and colorful character – delighted to shock and amuse the jury with his lack of pretense that he was anything other than a scoundrel. In all, he was a nervous, twitching combination of cringing self-depreciation and sly braggadocio with the vicious mark of Cain upon him. His story implicating Campbell and exonerating himself was, although implausible, well rehearsed. And he stuck to it as though his life depended on it – which it did.

But there remained a gaping hole in the prosecution's case. Not a single one of the prosecution's many witnesses had actually seen Campbell with the assassins. No one had seen him walking with Kerrigan and the shooters to McGeehan's saloon the first night, as Kerrigan claimed he had done. No one had seen him inside McGeehan's saloon that night. No one had seen the killers inside Campbell's Hotel, or going in or out of it, on either evening. No one had seen them departing from Campbell's on the morning of the murder.

This alone, it seemed to me, was enough to cast reasonable doubt upon the prosecution's case. Sure, what Kerrigan had actually done with the men – parading them around outside but not going in –

wasn't fully understandable without my testimony that Kerrigan was acting with the Pinkertons. But when the jury in a criminal case gets the impression that they aren't told the whole story, they should give the benefit of any doubts to the defendant. So in spite of the defense's mistake in not putting me on, I think that in front of an impartial jury, the prosecution's case was fatally flawed.[1]

And then McKenna took the stand. He couldn't directly corroborate Kerrigan's claims because he was in Tamaqua on the night before and on the morning of the Jones murder. He could only present his own claims, wholly uncorroborated by any other witness, that on various occasions Campbell had confessed to him. He was a moon-faced Irishman with high forehead, wide eyes, full mouth and dimpled cheeks that seemed always on the verge of amusement – a fit face for concealing the viciousness and treachery of a professional liar. He was sporting a handlebar mustache and wearing eyeglasses, which I know he didn't need.

Of course, he started out by rehashing all of his usual inflammatory and irrelevant blather about a region-wide Molly Maguire conspiracy. The presiding Judge had been handpicked by the coal companies. Finally, the prosecution asked McKenna to state any conversations which he or anyone else had ever had in the presence of Alexander Campbell in relation to the murder of John P. Jones.

He replied that the first had been on the evening of the 15th day of July, 1875, and:

> "A.   ...Campbell on that evening walked with me from his own house at Storm Hill to Summit Hill and on our way we talked of this murder. I asked him what he thought about it.
>
> Q. About which murder?
>
> A. The murder of Yost. Well, he said it was a very clean job.

---

1 Editor's Note: See Honorable John P. Lavelle, *The Hard Coal Docket*, pg. 65, for an account of the retrial of Alex Campbell in the Common Pleas Court for Carbon County before an unbiased jury in 1993, using testimony taken from the original trial transcripts. The verdict: Not Guilty.

He guessed they would not have bothered with it but it was done
on a trade.
Q. That was all?
A. That ended the conversation upon that particular day with
Campbell. On the fourth . . . (of August) . . ." [interruption –
pages 508-509]

Alex had been warned about this alleged moonlight stroll from
reading McKenna's testimony in the Yost case. He told us it had never
happened, although he had seen McKenna that night in a different
bar in Storm Hill, and the detective was quite drunk. I have since
been able to corroborate Campbell's version of that evening from the
detective's own records.[2]

Regardless, the discrepancy didn't seem worth making a fuss over
– at first. Campbell wouldn't have implicated himself even if he had
actually said he " . . . guessed they would not have bothered with it
(i.e. – murdering Yost) but it was done on a trade." He wouldn't have
been guessing about motives if they had been his own. The statement
was, in fact, exculpatory of Campbell.

But two days further into McKenna's testimony, his story changed.
He was still on direct examination. The prosecution was now being
conducted by my old mentor from the Dougherty case, Attorney
Hughes, who had been given a full-time job by Mr. Gowen to deprive
the defense of his services.

The wily old wordsmith asked McKenna what must be one of the
most egregiously leading questions of all time:

"Q. If you learned from Alexander Campbell how he procured
Hugh McGeehan and James Boyle to go to Tamaqua and kill Yost
in the way of a trade for the killing of John P. Jones afterward, will
you explain and say how he managed that?
Defense Counsel: Objected to!"

I leaned toward Mrs. Campbell and explained, "On direct
examination they aren't allowed to coach their witness by asking a

---

2 See Pinkerton Agency expenses for J. McF., July 15, 1875.

question which gives him the answer they want. This is ridiculously bad!"

But the Court shamelessly overruled the objection and permitted McKenna to answer:

"A. He told me on the 15th of July, 1875, that he would not have bothered with Officer Yost of Tamaqua, in killing him, only it was done on a trade . . ."

So there in one gulp McKenna changed his story of Campbell "guessing" about why "they" had shot Yost, to a version in which Campbell confesses why "he" had bothered to kill Yost.

I had trouble staying seated and keeping my voice to a whisper as I exclaimed to Mrs. Campbell, "Liar! What a lie! He's lying!"

"I know!" she groaned and silently bit her lip.

I kept my seat and watched the defense lawyers. They had no reaction. Not one of them picked up a pen to note the contradiction. Was it too obvious to require noting? I hoped so.

The testimony of McKenna was so full of such inconsistencies that it boggled my mind to read in the newspapers, day after day, the universal praise for how convincing he was. In fact, he was also rambling and confusing. In further answer to the proper question of the first day about all conversations with Campbell, he continued:

"A. On the fourth (of August 1875) I again met Campbell in Tamaqua; I believe we met in the Columbia House. In the course of our conversation I asked him as to how Hugh McGeehan was – as I had subsequently learned from a conversation with Campbell that Hugh McGeehan was one of the murderers of Yost, and I had been introduced to this man McGeehan in Campbell's saloon upon the 18th of July.

Q. You had been?

A. Yes, sir, by Campbell himself. Well, he informed me . . .

Q. Who informed you?

A. Campbell, that McGeehan and a man named Mulhall had been blackballed – could not obtain work around the Summit. They had got work or were about to get work – I believe he stated they had got

*work at Tuscarora and that he calculated for to start McGeehan
in a saloon for the clean job that he had done at Tamaqua; what
he wanted now was a few good men from Schuylkill to go to
Lansford and murder John P. Jones, the mining boss whom he
claimed was the instigation of preventing these men from
procuring work." (pages 509-510)*

This version of McKenna's alleged meeting with Campbell on
August 4th was different from the two versions he had come up with
in the Yost case. When he first told the story in the Yost case he said
that Campbell was delivering liquor to the Union House and it was
only after Campbell had departed that the proprietor of the Union
House gave him the name of McGeehan as the one who had shot
Yost. Later in the Yost case he changed the story to say that on August
4th it was Campbell who first told him about McGeehan in the
Union House and who then said he wanted men to shoot Jones. In
this third version of the encounter of August 4th McKenna claims it
took place not in Kerrigan's hangout, the Union House, but rather in
the Columbia House where McKenna always boarded when in
Tamaqua. Also, in this version he claims that this wasn't the first time
he learned the name of McGeehan, having already been given the
name by Campbell, "subsequently" in some conversation that he
never describes. It is frustrating that neither the defense lawyers nor
the prosecution lawyers ever forced McKenna to state his claims in
clear, chronological order. In any event, if McKenna had been
recalling actual facts, rather than fumbling around in fiction, the
location where such an important encounter took place would have
been indelibly fixed in his memory.

To anyone with only a few hours acquaintance with McKenna,
such as myself, there was one part of his story that was positively
ludicrous. He claimed that on the day before the 1875 Schuylkill
County A.O.H. Convention, he had a chance encounter with
Campbell who begged him to go to a meeting and get the men to kill
Jones. Campbell's reason for asking him to get the men was stated to
be as follows:

" . . . he could not depend upon this man Kerrigan to get the
men for the reason why he supposed Kerrigan would get drunk.
(pg. 513) . . . and . . . he was afraid Kerrigan would get too drunk
and he could not be depended upon." (pg. 549).

This is preposterous. It was McKenna who had the reputation of
being the biggest drunk in the A.O.H. He often drank himself
unconscious. Pinkerton doesn't deny it but seeks to excuse his
employee's excess. He writes (pp. 379-380) that on the evening of
July 15 his man "rolled himself" into Alex Campbell's saloon[3] and
bought drinks for the house.

" . . . of which the detective was compelled to inbibe fully his
share . . .(but) . . . his mind was so unduly excited, brain so highly
stimulated and alert that he might make no false step; speak no
suspicious word, the liquor he swallowed produced no more effect
upon his organism as so much water."

What the temperance ladies would say about this amazing ability
to keep his mind unmuddled by the alcohol, we need not wonder.
Even granting it some credence, it has no bearing upon the fact that
Campbell, like everyone else, considered McKenna to be an
unreliable drunk. That he would ask McKenna to get assassins at the
meeting because Kerrigan would get too drunk is like a man in
quicksand casting away both ends of his rope.

McKenna goes on to explain that although he did attend that
A.O.H. meeting, he made no attempt to get Campbell the men.
Nevertheless, a week later he learned that there was another fellow
who was out trying to get men for Kerrigan to use to kill Jones.
McKenna testified that at about ten o'clock the night of September 1
the Bodymaster of his Shenandoah Division, McAndrew, abruptly
appointed him and two others to " . . . go to Tamaqua and meet
Kerrigan and from there to go to Lansford . . .to shoot John P. Jones."
He said this order was given to him and the men without any
discussion whatsoever. He gave the impression that he had played no

---

3 He must not have read his man's expense report. See note 2.

role in the decision to send himself and the men or in selecting the men. (pp. 644-646). This story is hardly credible. McKenna was the leader of the gang; the technical Bodymaster who allegedly gave the order, was not the leader. This Bodymaster McAndrew was used by McKenna as a stage prop – a kind of *deus ex machina* – to order McKenna the Molly to do whatever McKenna the Detective should require.[4]

Be that as it may, the arrival of McKenna in Tamaqua with the two men to find Kerrigan and kill Jones gave rise to the destruction of McKenna's credibility as an honest detective.

He swore that they arrived at eight in the morning, and after dropping off his valise at his own room in the Columbia House, he took the men to the Union House. He was looking for Kerrigan but Kerrigan was not there. He does not say whether he inquired of the proprietress as to Kerrigan's whereabouts, although she was a confidant of Kerrigan's gang. It is permissible to infer that he at once realized that Kerrigan had gone to Lansford to kill Jones. He testified that he obtained a room from the proprietress and put his men to bed, and then:

"I left them there and returned to the Columbia House and I wrote out my reports and determined to send them to Mr. Franklin upon the two o'clock train in order to have a return from Mr. Franklin and to have Captain Linden on hand if I would be forced to kill John P. Jones and so I could save his life. I told them to wait until I would send for them at their homes which were at Gilberton." (pp. 583-584)

He also testified that he had sent his two men home " . . . on about – two o'clock along the cars going down towards Mahanoy City." (pg. 583). So everything – his dispatches and his men – were sent out at two o'clock in the afternoon.

---

4 See Author's Note – Appendix, as well as Captain Linden's contradictory testimony (p.654) that prior to September 1, he had instructed McKenna to make certain he would be one of the men selected from the Shenandoah Division to kill Jones.

Mr. McKenna's slip of the tongue about the possibility that he still might be forced to kill Jones is illuminating. His ostensible plan when he arrived in Tamaqua was to find Kerrigan and use the men to kill Jones. The planned assault had now become a mere "if", a mere possibility. He sent a report to Captain Linden to tell him so. This could only have been upon McKenna's assumption that Kerrigan had gone to do the job, relieving him of the need to do it. His subsequent protestations that he couldn't imagine where Kerrigan had gone until ten o'clock at night when the proprietor of the Union House informed him, and then it was too late to warn Jones, are without merit. He either knew or suspected where Kerrigan was since eight o'clock in the morning when he went to write up his reports. Possibly, Linden was prepared to save Jones' life should McKenna be required to supply the murderers, and he backed away upon learning that Kerrigan had gone to do the deed. In any event, the police were removed from protecting Jones on that very evening.[5]

The above-quoted statement of McKenna could also be construed to mean that he sent his two men on the two o'clock train to deliver his reports to Franklin and Linden. Linden was then in Shenandoah, through which the men would travel on their way home. If so, they would have known that McKenna was a detective.

McKenna wasn't grilled about this statement by the defense. Lawyers are leery about asking questions to which they don't know the answers. In this instance, however, there were no answers that would not have been destructive of McKenna's credibility.

Defense counsel were nevertheless fairly effective in showing that the detective did nothing to save the life of Mr. Jones. They established that by ten o'clock at night on September 2, McKenna was told by the hotel proprietor that Kerrigan had gone to kill Jones, and Jones lived only six or seven miles away:

> "*Q. Did you make any effort to save the life of Jones?*

---

5 See testimony of Charles Walton, Coal & Iron Police.

A It was not possible for me that night.

Q. Was Captain Linden in Tamaqua?

A. He was not.

Q. Were there any parties in Tamaqua that you knew?

A. There were several parties in Tamaqua that I knew.

Q. Any of the Coal & Iron Police there?

A. Not that I knew of.

Q. What did you do after you heard this? Did you make any effort to save his life?

A. I went and reported it: I had already made efforts to save his life.

Q. That night, after ten o'clock?

A. I was in company previous to this . . .

Q. That is not the question. Answer the question. Did you make any effort that night – In other words, what did you do after ten o'clock?

A. I found it impossible to make any effort that night in order to save my own life.

Q. What did you do?

A. I went to my hotel and wrote out my report and went to bed.

Q. In the morning at what time did you hear of the death of John P. Jones?

A. I should judge about seven o'clock or half past seven maybe. I can't be particular as to its date.

Q. Why would your life have been in danger?

A. How? I couldn't tell who would be watching me. Besides this, I had strict instructions from my superintendent to make reports only to him or Captain Linden. I kept up communication with no other authority and I was well aware that previous to that, Jones was on his guard and he was aware that he was going to be assassinated." (pages 587-588).

I could hardly contain myself when I heard this excuse and I exclaimed to Mrs. Campbell, "It's ridiculous! In one breath he claims that all the Tamaqua murderers trusted him so much they confessed their darkest crimes to him, and in the next breath he claims he was worried they were watching him! Which claim is the lie, or is it all lies?"

She shook her head in sad disbelief and said, "What's worse is his admission that he wouldn't report a murder attempt to anyone other than his Agency even if it should cost an innocent man his life! What credibility or moral authority can the jury find in him after that?!"

I nodded agreement and said, "He could have walked over and warned Jones or warned the Pinkerton men who were supposedly standing by to prevent the murder!"

"No such squad was standing by," she said. "Otherwise they'd be here testifying in this court."

"Obviously," I agreed.

On redirect examination the prosecuting attorney asked McKenna to describe the arrangements which had been made with Mr. Franklin and Captain Linden to intercept the murderers of Mr. Jones or to capture them and prevent the commission of the crime.

"A. We had an arrangement made to telegraph by cypher. Captain Linden prepared with his men to be ready upon learning the information. I was bound to find out what day was set to kill Jones. Not being fully aware as to where Captain Linden was located – I was not aware of where he was located – upon the second day of September at ten o'clock at night I could not dispatch to him and on leaving Carroll's (The Union House) I found I could not dispatch either to Mr. Franklin. The telegraph offices were closed." (page 601).

The notion that McKenna couldn't get a telegraph to Linden because he didn't know his exact location is absurd. The arrangement which had been made to telegraph Linden by cypher would have been farcically incompetent unless someone was stationed to receive the messages and get them to Linden wherever he might be. If McKenna had wanted to save Jones' life, he would have telegraphed Linden at eight o'clock in the morning when he arrived in Tamaqua and didn't find Kerrigan.

When I heard this testimony I exclaimed to Mrs. Campbell, "That man and Captain Linden and anyone else connected with their telegraph warning system are liars! We need my testimony to explain why they acted as they did!"

She replied through tears of frustration, "An' they're also murderers who are tryin' to hang Alex for what they did themselves!"

"We've got to stop them!" I said.

McKenna had a final broadside to fire on direct examination. He said he ran into Campbell in Tamaqua the very day after the Jones murder and the ever so trusting liquor dealer made admissions that totally clinched the prosecution's case against him.

"A. . . . . It was after the arrest of Kelly, Doyle and Kerrigan for the murder of John P. Jones. He seemed to be very much agitated about the arrest. He swore at Kerrigan and stated if Kerrigan had been any kind of an engineer he could have had these men at their homes in Mount Laffee before they were arrested. He stated it was the cleanest job that ever was done but the men were mislead by Kerrigan, and that he was going on the following day to Luzerne to try to raise money from the organization there for the purpose of the defense of those prisoners and he hoped that Schuylkill County would do their duty in this respect and if they did not that he himself was willing to take a mortgage upon his chattels or property to the amount of one thousand dollars and pay it out of his own pocket for the prisoners defense. He said as they had got into it now they must be taken out._

Q. Did he say at any time where he met Doyle and Kelly and Kerrigan?

A. He stated that on the night previous to the murder of Jones that these men stopped in his house and of course that he was of the opinion that there were parties there, sisters of the murdered man Jones, that lived adjacent to his house. Those parties would swear that those men had stopped with him but he said he could knock that higher than a kite because he would swear he had never seen the men; that they were never there at all, and did not

know Kelly and Doyle; that he knew Kerrigan very well and that
he would swear he had not stopped there.

    Q. That is what he would swear to?

    A. Yes, sir. He did not care about that." (pages 519-521)

After this fusillade of words, there was a pause and I whispered to
Mrs. Campbell, "Has McKenna forgotten his story that your husband
stopped Kerrigan from coming to murder Jones because the man was
so small that he'd be too easily identified and, therefore, he didn't
want Kerrigan on that job at all?"[6]

"I know!" she whispered back. "He's making out that Alex is a
complete idiot. Who could believe that the murder was the cleanest
job that ever was done after the two of them went parading around
town with a dwarf?!"

"Right," I agreed. "And only an idiot would feel safe that he could
overcome the testimony of Mr. Jones' two sisters by simply swearing
that he'd never seen the men!"

"Ach, it's all such foolishness!" Mrs. Campbell exclaimed.
"Anyway, we know from the Yost case that neither of those poor
sisters of Mr. Jones ever saw Kerrigan or his murderers going into our
place. No one did. 'Cause they weren't there!"

Court was recessed for the day shortly after this testimony and I
approached a young lawyer on the defense team, the only one of
them who would talk to me.

"This case boils down to whether the jury believes McKenna or
Alex Campbell," I told him. "You've got to put Campbell on the
stand. No one who listens to him for five minutes will believe he
could have uttered such drivel as McKenna just put into his mouth.
His very demeanor will prove McKenna is a liar!"

"I'll talk to my brother about that," he promised.

His brother was the lead defense counsel and I exclaimed my
satisfaction, "Thank ye! Thank ye very much!"

But an hour later outside the courthouse he told me, "We can't

---

6 Page 548.

put Campbell on the stand. He won't lie. He'll admit that he suspected a mine boss was in danger."

I nodded my head sadly and acknowledged, "This jury might hang him for that alone."

The young lawyer smiled encouragingly and said, "You'll feel better about it when you hear the defense witnesses which I've rounded up."

Actually, I felt better about Campbell's chances for acquittal while the prosecution was still winding up its own case with the testimony of Mr. Jones' sisters. Neither of the two who lived adjacent to the Campbell's, nor a third who often visited them, ever saw Kerrigan or his killers going in or out or nearby. And that wasn't because they weren't looking. One of them related that two weeks previous to his murder, their brother had been warned he was in danger, and a week previous to his murder his family had come to her house to sleep. Therefore, for weeks the sisters had been on the lookout for suspicious characters. But the only such character they ever saw around Campbell's place was one whom they saw several times – it was McKenna! (pages 294-299). This should have put a doubt into the hardest Modoc heart, whether the three killers had shuttled in and out of Campbell's place as Kerrigan was claiming.

But from my own point of view, the most revealing prosecution witness was Captain Linden. He testified that when John P. Jones was identified as the mine boss who was to be shot he notified members of the Tamaqua Vigilantes to warn Jones. He then told McKenna that while Jones' life was probably safe, McKenna should " . . . do the best he could to be selected as one of the men [to be sent to shoot Jones] because then we could be more certain to capture the whole business." (pg. 45). Therefore, McKenna " . . . was endeavoring to be selected so that the men should come from the Shenandoah Division." (pg. 659). He goes on to say:

" . . . on the first of September I received a dispatch from Mr. Franklin of Philadelphia stating that . . . a mining boss at Raven Run had been shot dead and . . . for me to proceed there immediately . . . "

*I left there immediately on the first train on the first of
September and that is why [McKenna] never knew where I went
. . . I was waiting for him to come down to Shenandoah.*

*Q. State whether it was contemplated that you would be there
[Tamaqua] on the second of September.*

*A. Yes, sir; the instruction was to remain until he came down
with his men.*

*Q. But you were called away by this telegram on the first of
September to investigate the murder. (at Raven Run).*

A. Yes, sir . . ." (pages 654-655).

This testimony certainly contradicts McKenna's pretense of being
unprepared to resist his Bodymaster's surprise announcement that he
and two men must go to assist Kerrigan in killing Jones. The whole
thing was planned. It corroborates the fact that McKenna and the
Pinkertons had been in control of everything that had been done by
the Shenandoah Division and – after the Yost murder – by the
Tamaqua Division.

I went to my contact on the defense team and pleaded, "You've got
to let me testify as to the Pinkertons' deal with Kerrigan. It explains
why he signalled the posse from where he and his men were hiding
and why they didn't immediately arrest him. It explains that
McKenna's bringing his own people to do the job on Jones was an
ultimatum to Kerrigan to go through with it himself. It explains why
Linden wasn't in Lansford where he should have been if he was really
standing by to save Jones' life. He was in Tamaqua encouraging
Kerrigan to get on with the job. On September 1, he would have
known that Kerrigan had gone to do it. That's why he felt it best not
even to wait in Tamaqua for McKenna and his boys to arrive, but to
go to Shenandoah to interrogate the Raven Run witnesses. It all fits
perfectly."

"It complicates the defense," my young lawyer friend objected.
"We'd rather keep it simple."

"But nothing else makes any sense," I persisted. "Why did
Kerrigan parade his men all over the place but not go into

Campbell's? Why would an honest detective take men to murder Jones and simply do nothing when he found that Kerrigan had gone to do it? Why wasn't Linden in Lansford to save Jones? Why were the Coal & Iron Police pulled away the night before his murder?"

The lawyer countered, "Those are all good questions. But standing alone they create sufficient doubt to permit the jury to find Campbell not guilty. We don't need you."

"Look at that jury, man!" I exclaimed. "They aren't going to give Campbell the benefit of any doubts. You're going to have to shock 'em with the truth!"

He scratched his head dubiously and said, "But to anyone who's been reading the newspapers, the truth is . . . somewhat . . . bizarre."

I couldn't argue with him about that – the truth being that the much-heralded Molly Maguires were, in fact, Gowen and his Pinkertons aided by Vigilantes and some Irish thugs cultivated in varying degrees of ignorance.

*   *   *

The defense attorneys were unwavering in their decision not to put me on the stand. A lawsuit is like a prizefight, with each contender fortified to take punishing blows. Without me, our side couldn't knock out the prosecution. But we conclusively demonstrated why it was that none of the prosecution's observers had ever seen Kerrigan or his men in or near Campbell's. Our witnesses unshakably testified that the killers were never there. Kerrigan had lied about that, plain and simple.

The prosecution was down to no more than the word of McKenna, that Campbell was a Molly and had confessed his crimes to him thinking that the detective also was a Molly. The defense counterpunched with a member of the A.O.H. who said it was a legitimate benevolent association, and there was no such organization as the Molly Maguires, and any violence had been due to McKenna. Two men, one of them a saloonkeeper in Mauch Chunk, said that McKenna had been a hopeless drunk. A number of character witnesses vouched for Campbell. Our lawyers thought that this would be sufficient to win a decision. And it should have been.

But what was bellowed out triumphantly by the jury foreman was, "On the charge of murder in the first degree – Guilty!"

Mrs. Campbell was almost incoherant with anger and with grief. She turned to McKenna sitting across the aisle and hissed that he was a "dirty little killer!"

Turning to the spectators, she asked in an agony of doubt, "Gowen and his hirelings aren't going to get away with this?!"

Then she threw herself sobbing into my arms.

Chapter 18

Gowen Triumphant

Campbell had been a prominent labor leader and politician. His conviction on the uncorroborated word of the two notorious characters who admittedly had perpetrated the murder themselves sent shock waves through the anthracite region. The Irish communities were dumbfounded and angry. Some among them sullenly suggested that the American Experiment wasn't working for them.

The other communities were worse affected. All of the anti-Irish, anti-Catholic sentiment, which welled in any of their hearts, found an outlet in the legal vengeance that Gowen had made respectable. Everywhere, voices clamored for the blood of more Mollies. Franklin B. Gowen and his Pinkertons were prepared to slake that thirst with stacked juries, handpicked judges, and Irish villains.

The next alleged Molly to be tried was a member of McKenna's own Shenandoah Division named Thomas Munley. He happened to be purchasing whiskey to medicate his sick child[1] when McKenna and the other Raven Run murderers burst into the saloon with a posse less than an hour behind them. McKenna bought drinks for all, and then they went their separate ways. But the defendant had seen too much. He had seen that Tom Hurley and McKenna himself had been part of the death squad.[2] Of course, the hapless fellow had to

---

1 Editors' Note: Whiskey was a common nostrum in those days.
2 Editors' Note: See H. T. Crown, *A Molly Maguire On Trial.* pp. 142-146

be silenced. The best way to do this was to hang him for the murder.

Only two of the many Raven Run witnesses came close to saying they had seen the defendant at the shooting, and their identification was shaky. The colliery owner said he had seen him and shot at him, but this was from two-hundred yards away. A woman could only say the defendant "looks like the man" she saw. The family of the accused testified that he had been home at the time of the murder. They were corroborated by several men from Raven Run who had seen the murderers, and who knew the defendant, and testified he wasn't one of them. This should have acquitted him. But he was a member of the A.O.H., and he was convicted on McKenna's word alone.

The next of Gowen's triumphs was the conviction of Jack Kehoe and his Bodymasters who had assembled on June 1, 1875, to plan the workingmen's demonstrations against the opening of the collieries. They were convicted of planning the McKenna gang's assault upon Bully Bill Thomas which McKenna himself engineered some twenty-seven days later.

I mention this case again because it was the first Molly case in which I was permitted to act as part of the defense team, and it led to an encounter with Mr. Gowen.

In its opening statement the prosecution made a great show of bringing the Pinkerton Agency's detective reports into the room and piling them upon the counsel's table. Their spokesman shook his finger at us and challenged, ". . . if Mr. [McKenna] tells anything upon this stand that is not true, he can easily be contradicted by his reports . . .!"[3]

This was highly improper. McKenna's reports were as susceptible to containing lies as was his testimony. But the Court allowed the piles of paper to sit there in mute confirmation of whatever he said. I was so annoyed by this prejudicial display that I decided to return to the courtroom on nights when I was free of other duties and copy

---

3 See Commonwealth v. Kehoe et al.,pg. 12

every word of the reports. To my surprise, the papers contained some of McKenna's expense accounts.

On the third night of this drudgery work I was interrupted by Mr. Gowen who came into the otherwise deserted courtroom.

"Ah, Mr. McWilliams! You're working late," he said, as he walked to the table to see what I was doing.

His demeanor was cordial enough, but I noted that he didn't offer to shake my hand. I stopped copying and looked up at him, but didn't extend my hand either.

"Yes, sir, it's late," I agreed. "I don't suppose there's anything here to help our case, but duty requires it."

"I would expect no less of you, my boy," he muttered as he examined the documents I wasn't then copying.

He asked some polite questions about my da' and ma' and our family, and I reciprocated by asking after his wife and charming daughter, and then he left.

The next evening I resumed my copying but found the expense accounts had been removed from the pile. I was relieved that I'd copied them first.

I believe the Bully Bill defendants were more ably represented than some of the others. We destroyed with pure logic the Pinkertons' claim that McKenna had been promised he would never have to testify. [4] Our Mr. L'Velle made an impassioned argument in favor of the labor movement that Gowen was attempting to destroy with his instigator McKenna:

"Immediately prior to 1873 while the Miners' and Laborers' Union was in the heyday of its prosperity in this county — I say the union, and God bless it for the good work it did in this community — there was not a transgression or serious crime of any character in our county for years . . . there was no such thing as a murder case in Schuylkill County, not until the emissary of death, James [McKenna], made his advent into this county and . . . I propose to

4. Commonwealth v. Kehoe, et.al., pg. 210.

prove to you that no crime has been perpetrated in Schuylkill County except that which he himself assisted to plot, to counsel, to perpetrate and to conceal afterward as far as he himself was individually concerned." [5]

But Mr. Gowen trumped us when the Court permitted him to testify scandalously to facts not in evidence in his closing argument. He testified that the men engaged in the workers' demonstrations against opening the mines, *"... claimed to be labor reformers ... but we now know [them] to be Molly McGuires! ... Look around this crowded courtroom ... Do you know why it is that ... no man lifts his hand against one of these prisoners? Do you not know that with the full knowledge that Jack Kehoe is a murderer ... and many more of these men are murderers, their lives are as safe ... as my own life is? Why is it? Because there is ... a well-founded belief that these men are in the custody of Justice, who will vindicate herself upon them through the instrumentality of your verdict! Will you then permit them to escape? Will you ... disappoint the righteous expectations of a whole Commonwealth and turn these men loose again upon society?"* [6]

This came close to threatening violence if the jury didn't follow public opinion and convict the defendants. Their verdict of Guilty was forecast by the newspapers and universally expected.[7]

I haven't the heart to go through all of the trials to compile a roster of the innocent and the near innocent from among the twenty lads who were hanged as Mollies. Quite as much of a scandal was that so many of the actual murderers never even faced justice – McKenna and his gang, Powderkeg Kerrigan, Linden and his Vigilantes, and of course Gowen.

Jack Kehoe was convicted of murder upon the flimsiest of evidence. McKenna had worked hard to tie Kehoe to a current murder, but he wasn't able to do it. So the Pinkertons went back to

---

5 Commonwealth v. Kehoe, et al.,pg. 193
6 Commonwealth v. Kehoe et.al.,pg. 191
7 See Author's Note in appendix

1862 when there'd been violent agitation over the unfairness of the Civil War draft. One of the men active in promoting the draft was a "ticket boss" in the mine where the Kehoe family worked, charged with docking the miners when the coal which they produced contained too much gob.

Jack Kehoe had been heard shouting at the man, "You son-of-a-bitch, I will kill you before long, because you are robbing me and robbing the men by your docking." This was three weeks before the man was murdered.

On the day of the murder the same boss conducted a pro-draft rally. Kehoe and other anti-draft men heckled him from the crowd and threatened to kill him for seeking to conscript poor miners who couldn't pay the $300.00 required to purchase an exemption.

That evening the ticket boss was waylaid and beaten by a gang of men. He seemed to recover but died two days later. He never identified Kehoe as one of the men who had beaten him. At the time there was thought not to be enough evidence to charge anyone with the crime. But things were different fourteen years later. Four Irishmen other than Kehoe were arrested and tried for the murder in 1876 and 1877. Two were convicted and two acquitted. In early 1877 Jack Kehoe also was brought to trial for the murder.

There was no witness who could identify Kehoe as one of those who had attacked the ticket boss. Two men testified that Kehoe was with them at the time the boss was beaten. Another swore that he saw Kehoe on a hotel porch at the very moment he heard the boss cry out for help. The prosecution's case was rife with reasonable doubt. But since Campbell's conviction the newspapers had been touting Kehoe as "King of the Mollies." He was convicted. In a last effort to save his life he applied for clemency and was able to present affidavits from the other two men convicted of the same crime. Both men admitted that they had killed the ticket boss and both swore that Kehoe had been in no way involved.[8] He was hanged anyway.[9] Even the anthracite region newspapers began to wonder about the sufficiency of evidence required to hang a Molly.[10]

I'll mention only one more scandalous murder conviction, that of our own Pat Hester from Locust Gap. He was the owner of the saloon in which me and da' and the family had plied McKenna with liquor while we convinced him of our fake train wreck story. Poor Hester had once been a noted leader of the A.O.H. But he had lost his influence in a fight with our pastor over whether one of his secret society members could be buried in St. Edward's Cemetery, which sat atop the mountain on which our house was built. Hester pushed aside the priest and buried his man after making forced entry. For this crime he served two-and-a-half years in prison, and he was demoted by the A.O.H. This was one of the reasons why the A.O.H. couldn't attract more members in the Shamokin area.

Anyway, the Pinkertons spread the word throughout all the jails in the anthracite region that the inmates could obtain their immediate release if they could help convict a Molly Maguire ringleader. This advertisement enticed from confinement the services of Pottsville's least favorite son, Kelly the Bum.

In his career the Bum had been guilty of murder, highway robbery, larceny, assault and battery, mayhem, rape, and the subornation of perjury. Pinkerton captures the essence of his informer with a horrible little vignette in his book (pg. 238). He tells that late one night the Bum and some boys of McKenna's Division broke into an unlicensed tavern run by an old woman. They robbed her of her money and drank up her wares. When she protested, the Bum picked up the woman and pressed her, face down, on an almost red-hot coal stove. He would have fried her to death had not one of his

---

8 Pardon Board Records, John Kehoe file.
9 Editor's Note: On June 21, 1980 the Governor of Pennsylvania on recommendation of the Board of Pardons, issued a posthumous pardon to John Kehoe.
10 *Shenandoah Herald*, April 11, 1878.

companions grabbed him and pulled him away. The old woman almost died. The Bum was never charged with this crime. Once again, McKenna could not report it without revealing his identity.

After being released from jail by the Pinkertons the Bum confessed that he and a gang of men had murdered a mining superintendent in the attempted robbery of a payroll eight years previous. He claimed that Pat Hester had suggested the robbery to him and his buddies. No one came forth to corroborate his story. Six witnesses contradicted his statements with regard to Hester's complicity in the robbery, and fourteen contradicted him on other significant points. His infamous character and his deal with the Pinkertons were fully proven.

Hester's family prepared a banquet for him on the evening of the day the verdict came down, to celebrate his expected acquittal. But they had to cancel it. The verdict came down, Guilty! The Bum was never charged with the murder, though he admitted pulling the trigger.

This was a shocking demonstration of the fact that Franklin B. Gowen wielded the power of life and death over any Irish labor agitator who might be foolish enough to cross him.

*   *   *

On June 21, 1877, known in the anthracite region as The Day Of The Rope, the first ten of the men were hanged, six of them at Pottsville and four at Mauch Chunk, including Alex Campbell. The nation's newspapers banner-lined the event and described with prurient intensity the blood lust and hysteria of the public spectacle stagecrafted by Gowen. All of the condemned – the guilty as well as the innocent and the near innocent – met their fates bravely. It was probably providential that there were some truly guilty men among the victims, for it was the recognition of a shared guilt among the Irish community which prevented them from exploding into violence, as the newspapers had been predicting that they would.

I need add only one thing of my own knowledge – what I saw

inside the Carbon County Jail where the hangman's scaffold had been set up in the wide corridor outside of Alex Campbell's cell.

Spectators crowded all of the ground floor space not occupied by the scaffold, and they stood on the stairway and the iron balcony giving access to the cells on the second story. Campbell had said goodbye to his family and had made his last confession. He sat stoically in his chains on his cot, showing no emotion, until the sheriff came to the door and told him, "It is time."

Campbell immediately got up and took a stride toward the sheriff, but then he stopped.

"I am innocent," he declared with a deep, steady voice.

He stooped and ground his right hand into the grime on the cell floor, then rose and pressed it against the wall of the cell, leaving a dark palm print.

"There is the proof of my words," he announced. "That mark of my innocence will never be wiped out!"

His final words, spoken on the scaffold, were more Christian than any I could have mustered then, or later.

"I forgive everybody," he said.

I have practiced law in the anthracite region for many years since — represented several men in the Carbon County Jail. The last time I looked into cell number seventeen, Alex Campbell's handprint was still there. By last report, it still is.[11]

*   *   *

About here is where I thought my manuscript would end. I would finish by saying something, perhaps quoting Edmund Burke, about the need for good men to fight evil. To that end I would publish my manuscript and condemn the murderous Gowen/Pinkerton conspiracy in the strongest of terms, and challenge

---

11 Editor's Note: The President Judge of the Court of Common Pleas for Carbon County attests that as of 1994, the handprint was still there in spite of attempts to remove it. Lavelle, *The Hard Coal Docket*.

them jointly or severally to sue me for libel. Since truth is a defense to libel, every word of my manuscript would be both sword and shield to me.

I worked on it as diligently as I could. But under the circumstances my progress was slow. I was a twenty-year-old law student with no money or prospects. I wouldn't have reminded Mr. Gowen of his offer to help me, even had I thought he'd still be willing. It took me another year to complete my studies and become admitted to the bar in 1879. Then I had to establish a law practice. As anyone who has done it will tell you, for about the first five years a young lawyer must work night and day merely to survive. Only the bad cases and worst clients come your way, and most legal issues are new to you and absorb all time and attention until a repertoire of knowledge and experience is developed. It is no time to be starting a manuscript. Nevertheless, I persisted to accumulate and organize my source materials and to keep myself aware of the movements of my targets.

Allan Pinkerton published his book from which I have been quoting as well as several others, dictating from his notes to as many as seven "ghost" writers. He became rich and famous as the world's greatest detective, but succumbed to a massive stroke in 1884.

McKenna, now known as McParlan, continued with the Pinkerton Agency, his duties taking him all over the country and down to South America. He became superintendent of the Agency's office in Denver, Colorado. Recently he ran afoul of the famous lawyer and my good friend, Clarence Darrow, who found him to be up to his old tricks again. Darrow proved that McKenna had gained access to the cell of a man imprisoned for the murder of a former Governor of Idaho, and bribed the man that he would set him free, as he had set free Kelly the Bum, in return for the killer's testimony that the assassination had been planned by the leaders of the miner's union. Darrow publicly exposed McKenna as a conniving liar and hypocrite, as we should have done when we had our chance at it.[12]

12 State of Idaho v. Haywood (1907)

Franklin B. Gowen had spent a ton of the Railroad's money to finance McKenna's undercover exploits, including the cost of all the Pinkerton personnel required to support him. But those expenditures, great as they were, were small in comparison with the monies Gowen had spent to buy up most of the coal lands in the anthracite region. He borrowed the funds from English bankers, and the Railroad's revenues were unable to cover even the interest on the debt without the use of imaginative accounting and ever more disastrous refinancings. The Reading went through several reorganizations. Finally, in 1883 Gowen was ousted from its management. He made a desperate bid to regain control in 1886, but failed. The Reading had been a kind of surrogate posterity for him – perhaps in replacement of his two sons who had died so early – but eventually he was forced to see it go, as well. However, he resiliently went back to the private practice of law in Philadelphia with his nephews as his partners, and they were doing quite well.

It was in the fall of 1889 that I girded myself for legal battle with Gowen, thinking my manuscript was completed. I had written and rewritten two drafts of it, satisfied with neither. Inexperienced with writing, I had permitted my narrative to stray and to contain so many names in its cast of actors that it was hopelessly confusing. Also, these versions were so filled with chasmed fears for the future of America that the simple truths of the story were obscured. But by September 1889 I had rendered the thing down to what I thought was presentable. It was no work of art. But it was sufficient, I thought, to reveal Gowen's murderous scheme to the public, and to cast at him my challenge to sue me if it was false. I made a copy of it with one of those new typewriters, containing upper and lower case letters.

In those days it was considered unsporting to publish such a thing without first giving the target of the story an opportunity to respond to its accusations. Therefore, I wrote to Mr. Gowen. I told him I had penned a version of the Molly Maguire saga that he might wish to challenge, and I offered to visit with him while he reviewed a copy of the manuscript, to benefit from his initial reactions, in which I was most interested. If he would not agree to this by the end of the year, I would consider myself free to publish without further consultation.[13]

There was no response for almost two months. Then in the last week of November, I received a dispatch as follows:

> *Will be at Wormley Hotel, Washington, D.C., from Wednesday, December 11 through Friday the 13th, and available in the evenings to review your project. A room is reserved which you may pay for upon arrival.*
>
> F. B. Gowen

The Molly Maguire conspiracy was spring-gunned to surprise yet another victim.

---

13 Editor's Note: For a book which marshals most of the facts and struggles to fit them into the Gowen/Pinkerton theories, see *The Molly Maguires*, by Wayne G. Broehl, Jr.

Chapter 19

Confrontation In The Capitol

I was not in an optimistic mood as I watched the treeless, black slag mountains of the anthracite fields slide past the windows of my train on the ride to confront Gowen in our nation's Capitol.

The death of the W.B.A. and the grisly executions of the Mollies had failed to bring peace. A severe economic depression reduced the Railroad's and the coal companies' profits and caused them to lower wages. The miners resisted with a series of violent, uncoordinated strikes. Fearing a resurgence of Molly Maguireism, the anthracite communities dealt ruthlessly with the workers. When a mob descended upon the Reading's way station in Shamokin, a posse of Vigilantes fired into them. In Scranton, Pennsylvania, the coal company's superintendent, William Scranton, ordered his private police to "shoot low and shoot to kill . . ." the marching strikers. Poverty and hunger descended upon the entire region.

Defying the threat from Gowen's hangmen, the Irish miners reorganized under the banner of The Knights Of Labor. Predictably, they were attacked by Gowen who claimed they were terrorists and that there was a "gang" within the union similar to the Mollies. This time we published the identities of our leaders and challenged the District Attorneys and the Judges to arrest them and try them in court. Of course, they could not, and they did not. This helped convince the people and, more importantly, the Catholic clergy, that the union was respectable.

But the Knights were mostly Irish. What was needed was a union which the Welsh, the Germans, the Italians and the Eastern European miners could embrace. So the Amalgamated Miners' Association was formed to attract all the nationalities. It was widely successful and it formed a coordinating committee with the Knights. This putting aside of ethnic differences was the most encouraging social development in the United States since the end of slavery.

Unfortunately, the railroads and the coal operators remained even better organized. They convinced the anthracite region's merchants and businessmen to support their pooled control over production, prices and wages and to resist the workers' demands. Once again the unions were forced to strike before they were prepared. By the time I boarded the train for Washington, they had been soundly defeated. Coal in the ground will last forever, while life loses its value as a man watches his family starve.

From the train station it was only a stretch of the legs to Washington's finest hotel, The Wormley. As I carried my bags into its four-storied elegance I saw it was more opulent than anything I had ever experienced, or could comfortably afford. It had a doorman. By heavens, it also had a booth in the lobby containing one of the new Bell telephones! Just as impressive was the gleaming brass and mahogany elevator going up alongside the red-carpeted staircase. What a wonder! I later learned there wasn't anything to compare with the Wormley in the entire city. Its restaurant was world famous for its Chesapeake Bay seafood and its turtle soup. It was the haven of the rich and powerful visiting Capitol city. I felt hopelessly out of place.

I went to the reception desk and told the man that Mr. Gowen had reserved a room for me. When he told me the price of the room, I gagged in dismay.

"There's been a mistake," I told the clerk. "May I speak to whoever is in charge?"

"Yes, sir," he responded. "That would be the owner, Mr. Wormley himself. One moment please."

He knocked on a door and, receiving leave to open it, he stepped inside and I heard him say, "Mr. Gowen's associate has arrived, sir. He has a problem . . . Very well."

He ushered me into the office and introduced me, "Attorney Matthew McWilliams, this is Mr. James Wormley."

Once again I was shocked. The owner, about my own age, was a large and handsome Negro. I shook his hand and he invited, "Please have a seat, Attorney McWilliams. How may I help you?"

He hadn't a trace of an accent. He was impeccably dressed in a conservative blue suit and he was obviously quite comfortable seated at his desk in his elegantly appointed office, and very much in charge. On the wall behind him hung a plaque that certified that James T. Wormley had graduated from the Howard University School of Pharmacy.

"Well sir, I'm embarrassed," I told him. "Mr. Gowen reserved the room, but now I find I couldn't afford to pay for more than one night. I didn't know that. With your permission, I'll go out an' look for more . . . ah, appropriate accommodations."

Mr. Wormley smiled, his white teeth flashing brightly against his black skin. I had the suspicion that I wasn't the first bumpkin whom Gowen had impressed with his wealth and power by reserving him such an expensive room.

"I'm certain Mr. Gowen would prefer that you stay here," he said. "He's not yet arrived but he's made arrangements to take you to supper with him tonight. Permit us to solve your problem. He reserved you a luxury room on the second floor, close to him. We aren't very busy this time of year. We'll be able to move you to the fourth floor at a rate that will allow you to stay three days. Would that be satisfactory?"

"Yes, it would, Mr. Wormley!" I exclaimed, much relieved. I rose and shook his hand and added, "I'm beholdin' to yer courtesy sir. Very much."

"Not at all, Attorney," he smiled again. "We are pleased to be of service."

I was excited to take my first ride in an elevator. The operator seemed very important. A bellboy escorted me to my room on the fourth floor, unlocked it and handed me the key, and showed me in. It was beautiful – plushly carpeted, with a bed, a wardrobe, a dresser, and a chair and desk on which there was an ornate oil lamp. It even had its own private bathroom, and there were two large windows in dormer recesses. I'd never been in anything so grand.

But the bellboy was distressed because the bed had been stripped of sheets but was not yet made up.

"I'm sorry about this, sir," he said. "The maid on this floor has taken ill. The second floor maid has been trying to keep up with both floors."

He went out into the corridor and in a loud stage whisper called, "Miss O'Neal! Miss O'Neal!"

The maid, Miss O'Neal, soon appeared in the doorway. She was in her late twenties and very pretty, with auburn hair and green eyes, though a bit on the slim side of the current mode.

"Oh, I'm sorry sir," she said to me, genuinely upset. "I hadn't thought this room was to be occupied today. I'm afraid 'twill take me some time to make it proper."

"Don't worry," I told her. "I'll leave my things here and take a walk around town before it gets dark. Ye'll have plenty of time to do it."

"That's kind of ye, sir," she said with a curtsy. "If ye don't mind, leave yer door unlocked. Your belongings are quite safe in the Wormley."

"All right," I agreed. She left.

I unpacked, placing my shirt and things in the dresser and my bag in the wardrobe. Sitting at the desk I took my manuscript out of my valise to check that it had traveled safely. It was then about a hundred-eighty pages and the title page accurately described its contents – *The Unrequited Crimes Of Franklin B. Gowen*. I laid it carefully upon the desk and put the empty valise into the wardrobe.

There was a knock on the door and Miss O'Neal called out, "Maid."

"Come in," I answered.

She entered and said, "I can finish now sir, in about an hour."

"I'll be gone longer than that," I told her. "No hurry."

I went into the bathroom, and when I came out Miss O'Neal was dry mopping the uncarpeted perimeter of hardwood flooring near the desk.

I put on my overcoat to leave and she pointed to the manuscript and asked, "Excuse me, sir, but is this about the famous Mr. Gowen of the Reading Railroad?"

"Yes it is, Miss," I answered. "Why d'ye ask?"

"Oh, sir, I'm originally from around Pottsville where all those men were hanged. From St. Clair, more exactly. D'ye know the place?"

"Yes, indeed," I answered. "I'm from Shamokin."

"I miss the coal region, the mountains an' everything," she said. "It's less hectic than the big city. I'm longin' to return some day. Me an' my daughter."

"It's beautiful where the mountains aren't scarred by the mining," I noted.

"Yes, well good day to ye, sir," she said as I left the room.

That was about three o'clock in the afternoon. Sunset was at four-thirty and it got dark at five. I got back from my walking tour at about six to find the door to my room ajar and a light on inside. I rapped before entering and found Miss O'Neal still there. She was polishing the lit oil lamp on the desk.

"Oh sir, I've just finished yer room," she curtsied. "May it be to yer liking."

"I'm sure it is, Miss O'Neal," I said as she scurried past me out the door.

I went to my manuscript and examined it and checked my clothes and toiletries. Everything was as I had left it. The room was spotless.

I took the manuscript to Mr. Gowen by walking two flights down the fire exit that was close to his room, number 57. The desk clerk had told me the room number, and that Gowen had arrived and was expecting me.

I knocked.

He came to the door in his shirtsleeves and shook my hand and said, "McWilliams. How nice to see you again. You've brought your paper with you. Good. Good."

"Hello, Mr. Gowen," I replied.

He stood in the doorway blocking my entry and said, "I'm only just arrived. Supper's between seven and nine. If you'll meet me in the dining room at eight, I'll have a couple of hours to read your work. It will give us more to talk about."

"That will be fine, sir," I replied, handing him the manuscript.

He closed the door. I noticed he didn't lock it. No one locked their doors when they were inside their rooms in the Wormley.

Supper with Mr. Gowen was an experience. I arrived at the chandeliered dining room first and sat at our table in wonderment. The place was appointed like a potentate's palace with expensive draperies, tapestries and paintings, white linen, gleaming silver and crystal, plush armchairs and monogrammed napkins. The waiters and busboys hovered so like flies around a jam jar that you wanted to shoo them away.

Mr. Gowen didn't offer to shake my hand again when he arrived at the table. He sat down and growled, "To begin with, your title is libelous nonsense!"

"Not if the contents prove it to be true," I countered.

The waiter came and took our order. After he left Gowen tried a different gambit, the laying on of guilt.

"When I invited you into my home I didn't expect you to be an eavesdropper," he said in reference to my overhearing his instructions to Pinkerton and McKenna.

"It was totally unintended," I replied. "Anyway, yer complaint doesn't fit well into the mouth of someone who packed our union with his spies."

He snickered guiltily and admitted, "I suppose I must concede that point."

Dinner was served and it was excellent, but I don't recall what we had – some kind of fish I think.

In the course of it I asked, "How much of the manuscript have ye read?"

"Past the place where you confronted McKenna with the fact that he was a detective, and into the Bully Bill affair," he answered.

"That's fast reading," I noted and asked, "Did McKenna ever report it to ye that we had found him out?"

He regarded me with suspicion that turned to smug amusement as he answered. "Wouldn't you like to know! That, my boy, must remain privileged information."

"Captain Linden must have figured out eventually that we were feeding him false alarms," I persisted.

"From the beginning the Captain knew the union was keeping an eye on him," he admitted.

"That's why, in Shenandoah, he and McKenna pretended they had a past acquaintanceship. So they could be seen together," I suggested.

He didn't respond to this one way or the other and was silent for a while, seeming deep in thought.

Eventually he said, "You people claim that McKenna would have taken money to convict his own grandmother. But you're wrong. He's a bird of a strange feather, to be sure. But he would never say a word against the union. There, he wouldn't budge."

After supper we sat in easy chairs in the lounge and had cigars. We exchanged pleasantries about the fine meal and the wonders of the Wormley Hotel, especially its modern conveniences.

Finally, Gowen asked rhetorically, "You know don't you McWilliams, that your little book doesn't matter. The issues you raise have been decided by Judge and Jury and their decision will always be accepted as the final proof."

"Ye're wrong, Mr. Gowen," I replied. "The first of the defendants were guilty as sin, regardless of whether yourself and McKenna were equally guilty, or whether the Molly Maguires ever existed. Ye were brilliant to argue those issues in the cases where they didn't matter. There were no counter arguments, and the guilty verdicts helped convince the public that those issues had been settled. It was evilly clever of ye, with emphasis on the evil part."

His face reddened with anger and I hastened to add, "But that was a public relations victory, not a legal decision. By the time those issues became relevant, yer compliant judges and predisposed juries did the rest."

He snarled and threatened, "I have powerful friends who approved of everything I did. If you publish that thing you'll never practice law in the coal regions afterward!"

"Perhaps that's true," I admitted. "An' perhaps if ye'd been satisfied with hanging only the guilty, I wouldn't risk it. But ye weren't satisfied. Ye had to make it seem that the workers' leaders were guilty. An' many of them were innocent, as ye knew damned well! So unless ye convince me that I'm wrong I'm going to risk it."

Gowen puffed on his cigar, rose to his feet, and before striding out of the lounge, he growled, "I'll get back to you when I finish it."

CHAPTER 20

GOWEN'S MYSTERIOUS DEATH

I slept late the next morning. When I went downstairs for breakfast the front desk had a message for me from Gowen. It said he would be occupied all that day and evening but would finish the manuscript by four o'clock on Friday the thirteenth, and if I was still in town I could call at his room at that time.

He had told me he would be engaged that morning in a hearing before the Interstate Commerce Commission, so I wasn't surprised. However, I was concerned by his suggestion that I might leave town without seeing him. After dinner[1] I wrote him a note that our meeting in his room on the thirteenth should be considered a definite and mutual commitment.

On my way out to visit the new monument to George Washington that had opened only the year previous, I stopped by Gowen's room on the chance he might have returned.

I knocked on the door and I heard a voice cry out, "Oh!" and there was a thump as though a chair had been turned over.

I tried the bedroom door and it was locked. I waited.

After a half a minute the door into the hallway from Gowen's bathroom opened. The maid, Miss O'Neal, looked out to regard me with dismay.

"Oh, hello sir," she said, looking and sounding very flustered. "Ah . . . Mr. Gowen isn't in . . . ah, I'm tidying up his rooms."

---

1 Editors Note: In those days "dinner" was what we now call "lunch."

"Sorry to disturb ye Miss," I said. "I've a note for him which I'll take down to the desk."

"Yes, sir," she said and quickly ducked back into the bathroom.

There was something suspicious about the encounter. I wasn't familiar enough with the practices of the Wormley to be certain what it was. The ungallant notion crossed my mind that Miss O'Neal might know Mr. Gowen more intimately than she had let on. But I immediately rebuked myself for the thought and went about my own proper business.

After enjoying the view from the top of the monument, I walked to the Interstate Commerce Building with the idea that I might observe the end of Mr. Gowen's performance there. I arrived to find Mr. Gowen already leaving with his client, walking toward our hotel. Along the way they met with another man and he and Gowen went into a different hotel. Not wishing to interrupt him, I went back to my room.

Hoping to engage Gowen again I went down to the lobby at seven o'clock and sat where I could unobtrusively observe the entrance to the dining room. I was there only a minute when Gowen disembarked from the elevator and strode purposefully out the front door. I wasn't wearing my overcoat, but I wanted to see if he would comment about the book, so I decided to go after him anyway. My indecision had given him a two-block head start. The evening wasn't bitter cold, but cold enough for me in my suit coat. I was relieved when he went into a store, but when I got there and looked through the window I took pause. It was a gun store. Gowen studied the merchandise for quite a while and then pointed at a pearl-handled revolver that a clerk demonstrated for him. He was purchasing a pistol! I reconsidered the advisability of accosting him alone on a dark street, and ran back to the Wormley.

I admit to being somewhat apprehensive. I was accusing Gowen of cold-blooded murder, or close to it. He was perfectly capable of pulling the trigger on me himself, in my opinion. My manuscript would permit him to infer, from his warped point of view, that me and my family were more destructive Molly Maguires than most of those he had managed to hang. The breakers burned in Shamokin had heated the hatreds that had consumed his soul. I didn't want to be his next victim.

It was with trepidation that I approached the door to his room at four o'clock the next afternoon. I was assessing the possibility that Gowen might think he could get away with shooting me. The doors to the rooms in the Wormley were frequently banged shut – by the occupants and by draughts of wind – so he might believe that a pistol shot could go unnoticed.

I took a deep breath to steady myself and rapped boldly upon the door.

There was no answer.

"Has he decided to break our appointment?" I muttered to myself and knocked again.

Again no answer. I was beginning to walk away when I heard him call out, "The maid is in here! Come in, Mr. McWilliams!"

It was strange that he had announced the maid. I opened the door and stepped inside carefully. Gowen was seated at the table opposite the fireplace, fully dressed in his suit. His bed had been made up and the room looked clean. Nothing seemed amiss and I closed the door. Miss O'Neal was standing to my right with her hands behind her. She nodded politely. I nodded in return and took three steps toward Gowen and picked up my book from his table. He continued to stare at something behind me with such frustration and anger on his face that I instinctively turned around. Miss O'Neal had followed me. She was pointing Gowen's new pearl-handled Smith & Wesson 38 at my nose, so close I could see the barrel was clean!

"What's going on?" I exclaimed.

"I came in and caught her reading your manuscript," Gowen explained. "She grabbed my pistol and made me sit here while she finished it."

"Now that ye're done ye can put down the gun," I said to Miss O'Neal.

"Sit down!" she ordered and waved the pistol toward the bed.

I sat there holding my manuscript and she said, "Good. Now I want ye to tell me something, the two of ye being lawyers."

"What is it ye want to know?" I asked.

"I want to know if ye've enough evidence in that book to convict this dirty killer in a court of law."

"No, not in a court of law," I shook my head.

"I was afraid not," she said and asked, "But what if he signs a paper that it's all true?"

"That would be under duress. It wouldn't matter," I answered.

"But ye do intend to publish it?" she demanded.

"Yes, unless he can convince me that anything I've put in it is false."

"Well, is there, Mr. Gowen?" She swung the pistol upon him.

Gowen was furious. He could hardly control himself as he retorted, "Go to . . . I don't owe either of you any explanations!"

"Oh, ye don't?!" Miss O'Neal discharged her words with equal fury. "Ye don't think so, ye lying cur! Perhaps ye should wonder how the likes o' me becomes a maid and remains a Miss with a half-grown daughter?!"

"I couldn't care less!" Gowen replied.

"Of course ye don't care!" she spat out her agreement with Gowen and turned to me with her story.

"Sure, it was me own fault what we did before we were married. I admit it. But Mick and I were pledged to each other an' we were so much in love . . . Anyway, it was nothing more than a thousand other couples have easily cured with an understandin' priest and a quick wedding. The day was all set. Then along came this one's damned Coal & Iron Pinkertons roundin' up all the A.O.H. they could lay hands on. Mick hadn't done anything, but he had to run. After readin' yer book I'm glad he ran. He went out to the Colorado gold mines an' he was killed. He might've been killed in Pottsville, to be sure, but there we would've been married. I wouldn't have had to come to the big city and been a scandal to my entire family."

She turned to Gowen and hissed, "That's why ye owe me an explanation – because ye destroyed me life with me family, ye connivin' pig!"

Gowen ground his teeth in silence as he considered her accusation. Then he rejected her with a dismissive wave of the hand, "As you say, it's your own fault."

Tears began to well in her beautiful eyes and she pleaded with him to understand, "Ye have no idea what our lives have been like in this hellhole of a city!"

Gowen reacted with hatred. He wanted to lash out at her, to punish her for her words and for his own guilt. His voice was cruelly deliberate as he said, "You might want to fix yourself up a bit . . . start yourself up in an older and more lucrative profession . . . to which you're well suited by your own admission."

"That's foul!" I exclaimed as I saw the pain coursing through Miss O'Neal's face.

She was left-handed holding the pistol. With her right hand she reached up to staunch her flow of tears.

Gowen saw this as his chance to strike at her physically. With his right hand he turned and grabbed the heavy lamp to his left upon the table. He was about to swing it back-handed into Miss O'Neal when she saw him. She put out both her hands to protect herself and the gun went off. The bullet entered Gowen's head behind the right ear and blew his brains out.

I was dumbfounded. Gowen's blood and tissue were all over the fireplace and the walls. He was no doubt dead. Miss O'Neal backed away from the horror, against the bathroom wall, and trembled in shock.

Out in the corridor James T. Wormley had been looking to find out why his second floor maid had not reported the progress of her rounds. He heard the retort and entered the room and saw everything.

"My God! My God!" he exclaimed. Pointing to the gun in Miss O'Neal's quivering hand he cried to her, "What have you done?!

What have you done?!"

She moved toward him with the pistol between her outstretched palms as though praying for his forgiveness. She opened her mouth to plead with him, but nothing came out. Seeing Gowen's body bent over the table pouring out blood, she gasped and cast away the pistol. It clattered on the table and slid under the corpse. She retreated back to the wall and buried her face in the corner where a large bureau stood out from it.

Wormley went gingerly to the corpse and felt for a pulse. That was the pharmacist in him, no doubt. He picked up the lamp from the floor where it had fallen and lit it and placed it on the table. Then he pulled down the blinds and closed the curtains of both windows.

Going over to Miss O'Neal, he pulled her out of her corner and holding her shoulders at arms-length demanded, "Why have you done this to us?"

It was a strange question, I thought.

Miss O'Neal looked him in the eyes and pleaded, "I didn't mean to do it. I didn't. I only wanted to make him to sign on to the book, is all . . . that it was true!"

"You're making no sense girl!" Wormley cried in frustration.

To me, he explained, "Our family fortune is tied up in this hotel. We'll all be ruined. Ruined!"

"It's not your fault," I tried to comfort him. "It's Gowen's fault. In so many ways, it's Gowen's!"

"His?!" he exclaimed incredulously. "How could it be his fault?"

I sensed that he was verging on hysteria, so I tried to keep my tone calm as I replied, "Listen, Mr. Wormley, and I'll tell it to ye. There's a lot of background to what happened, but the short of it is this. Gowen was about to brain the girl with that heavy lamp. She flinched and shot him. It was an accident . . . at the worst, she shot him in self-defense. In any event, it's not a capital offense. I'll swear to that."

Wormley had calmed himself and listened carefully, and for a moment he considered what I'd said.

Finally, he shook his head dejectedly and groaned, "That won't matter. We'll be ruined. Totally!"

"Why?" I demanded. "You're not to blame!"

"Don't be stupid," he implored. "Look at me. We're Negroes. They won't let us live this down."

To Miss O'Neal he agonized, "Is this how you repay us for our kindness, girl?"

"No, sir!" she sobbed.

"All right, let's all calm down," I interjected. "You both have to stop imagining what people might say and think and start analyzing this for what it is. It's a terrible accident, to be sure, and arguably a crime; but manslaughter at most. Don't let your fears turn it into anything worse."

Wormley threw up his hands and asked, "What can we do?"

"Do what ye'd do in the event of any other kind of accident," I told him. "Get on that telephone downstairs and call Gowen's law office and talk to one of his nephews. Ask his family how they want to handle this. Don't let them find out about it in tomorrow's newspapers."

Wormley considered this and mumbled, "I suppose . . . that's what we should do . . . of course."

"Let's get the girl out of here first," I suggested. "Then I'll go with ye to make the call."

We led Miss O'Neal up the stairs of the fire exit to my room and laid her on the bed with a wet towel to soothe her flushed face and forehead.

As we were leaving I told her, "Don't move nor utter a sound until we return, Miss."

*   *   *

It was about five o'clock when Mr. Wormley telephoned Gowen's law firm and asked to speak to his nephew and law partner, Francis I. Gowen.

I could hear only Wormley's half of the conversation. After expressing his regrets to perform such a sad duty, he came right out and said that Mr. Gowen had been shot dead in his room by a woman. There was a long pause and he had to repeat his statement twice.

Then he said, "I can't say exactly what the woman was doing in his room. I was too upset to ask her. You can ask her."

"A street walker? Here we call them hookers. That may be one of her occupations. Anything is possible."

"She says the gun went off by accident. I have no reason not to believe her."

"Because there was an associate of Mr. Gowen's who saw the whole thing. Yes. Yes. He's staying here. Mr. Gowen made the reservation for him."

"I won't. I won't do anything. Very well. I'll wait."

Mr. Wormley placed the listening tube back in its cradle and said to me, "He wants us to do and say nothing until he calls us back. I'm to keep you in your room and not let you talk to anyone. Same with Miss O'Neal."

We sat together in the lobby to prevent anyone else from using the telephone. We had nothing to say to each other in public. In less than two hours the telephone rang and Mr. Wormley snatched it up.

"This is Mr. Wormley. Hello, Mr. Gowen. We did exactly as you wished."

He listened intently for a long time and then he said, "Yes, I've heard of him. Who hasn't? I see. I see. Yes. I think that that will be best, under the circumstances. Goodbye, sir."

He seemed relieved. The lobby was empty of guests and he whispered to me, "They want us to make it look like a suicide until they can get here and deal with it themselves. I'm to pretend to discover the body at noon tomorrow. They'll arrange to have a news reporter on hand who will support the suicide story if we can make it at all plausible."

"I'm not sure I like that idea," I protested. "In fact, I'm certain I don't. Why the hell not tell the truth?"

"Let's get out of the lobby to talk about it," he suggested.

Back in Gowen's room the scene seemed even more grisly as it assailed the senses a second time. It was difficult to think clearly, but I tried to collect information.

"Mr. Wormley," I said. "From what I heard ye permitted the Gowen's to think that Miss O'Neal might be a prostitute. D'ye have any reason to believe that she is?"

"Of course not," he avowed. "We'd have fired her if we'd had the faintest inkling of that. I didn't rule it out when they asked because if they're afraid she might be a hooker they'd want to hush up the whole affair. That is what I want. That is precisely what I want."

He was a powerful man. He lifted Gowen's body away from the desk and dragged it to lie on the floor, face up, with the feet closest to the fireplace. Then he put the bloody pistol near the right hand of the corpse.

"It will look as though he stood in front of the fireplace mirror and shot himself," Wormley explained. "We'll lock the door from the inside with the key in it and go out by the bathroom door which locks itself. They can claim he locked himself in and committed suicide."

"Who all are the 'they' who are coming tomorrow?" I wanted to know.

He answered, "Mr. Gowen's two nephews who are his law partners, and also Captain Linden of the Pinkertons."

"Oh!" I exclaimed in surprise. "Now I understand all this duplicity. It's all Linden's idea. He has a knack for keeping his options open. You didn't give them my name, from what I overheard?"

"No, but their secretary called us to make your reservation," he said. "They'll probably have discovered your name by the time they get here."

A suspicion crossed my mind and I said, "Mr. Wormley, they might like nothing better than to accuse me of this shooting."

"That's absurd," he protested. "Gowen didn't fear you. Dozens of people saw you having supper with him Wednesday evening and smoking with him in the lounge. Besides, if it comes to that I'll testify that you didn't do it. I wouldn't have that on my conscience, believe me."

"I believe ye," I said. "But also I must caution ye now. I won't lie. If my name is revealed an' I have to answer questions to the authorities or to anyone else, I'll tell the truth. The whole story as it actually happened. When Captain Linden gets here ye'd best let him know that, first thing."

"Fair enough," he agreed. "But then you must stay away from the news people, stay in your room at all times after I discover the body. That will be at noon. I'll have your meals sent in and whatever else you require – books, newspapers, anything. We'll pay for it all. I'll report to you what's happening."

"All right," I agreed. "An' let's not forget Miss O'Neal who we've left alone for too long."

We went up to my room where Miss O'Neal lay on my bed. She hadn't moved since we'd left her. Mr. Wormley ordered her to go directly home and to return an hour early in the morning when he would rehearse with her what he wanted her to do and say regarding discovery of the body.

She seemed in a daze. It was clear that she was still very much agitated. Putting on my overcoat, I offered, "I'll walk ye home."

"I wouldn't advise it," Mr. Wormley cautioned me. "She lives in Swampoodle."

"What's that?" I asked.

"It's where the Irish live," he explained. "The worst slum in the city. Very dangerous."

"I don't come from such a fancy neighborhood myself," I dismissed his concern.

"You don't understand, Attorney," he said. "Even the police won't go there, except in the daytime, and then only in squads to pull dead bodies out of the creek."

Miss O'Neal stood in a kind of trance, shivering from the shock of what she had done, and I decided, "I've got to risk it. D'ye have a cane I could borrow?"

He loaned me a cane and I walked Miss O'Neal to her home. She wore a threadbare coat and a kerchief on her head and carried a sack containing an empty wooden cheese box for a lunch pail and a lidded tin.

At a grocer's she explained, "I must make supper."

We went in and she selected a crust of stale bread and had her tin filled with milk. When I added an entire ring of spiced bologna and paid for everything, she bit her lip and said nothing. The look she gave me was of shame and gratitude in equal portions.

The Swampoodle Slum, about a mile-and-a-half from the hotel, was on the far side of a newly erected baseball park. It was a filthy, crowded amalgam of hovels of every description. Sinister looking thugs hung out on street corners and in front of the numerous shebeen houses.[2]

The stench of offal, excrement and vomit mingled with cooking cabbage. She took me down an alley that was packed with mostly one-room shacks. Had there been no moon I wouldn't have been able to make my way.

The door to her shack was on the side and I could see into an unfenced area where there were rows of outhouses, piles of garbage, nondescript rubble and a populous, writhing cluster of aggressive rats. I shuddered. The coal region patches weren't fancy, but the houses usually had fenced yards and, though not always tidy, they were generally clean.

The inside of her hovel was about fifteen-feet-square. To the left of the door was a simple iron stove, and next to that a bare wooden table squatted, with no chairs. Light came from a bowl of oil on the table in which a burning wick floated. If the thing ever spilled, the shack would go up in flames.

To the right of this kitchen space the room was partitioned into two separate living areas by sheets that had been tacked into the ceiling boards. In the area closest to the door lived a family of four. The wife was plying her trade, sitting on a stool and knitting lace by

---

2 Editor's Note: Unlicensed taverns.

the light of the burning wick. Her husband was playing with an infant and a small boy on a feather bed. Incredibly, they seemed happy.

I said "Good evening" in Gaelic and the adults responded in kind. All four stared at me in wonder.

Behind the hanging sheets in the area remaining there was another bed and two stools. A thin waif of a girl, about twelve years, wrapped in a shawl, was sitting on one of the stools pretending to spoon feed a pathetically homemade rag doll which she had perched on the other stool. When she saw that her mother had come home her sunken eyes grew bright and she was very beautiful. They hugged each other.

"This is my daughter," Miss O'Neal introduced us. "Molly, this is a gentleman who purchased supper for us tonight, Mr. McWilliams."

"We're most grateful to you, sir," she curtsied and spoke like a child who had attended sister school.

After exchanging pleasantries I reminded Miss O'Neal to speak to no one and to call upon me if I could be of any assistance. Then I left.

I was in a hurry to be out of that place, afraid its innocent occupants would see their squalor through my eyes.

*   *   *

The next morning, Saturday, I went down to breakfast and then out to purchase some books and take a walk, being sure to return by noon. Mr. Wormley reported to me in the late afternoon and promised to have the evening newspapers delivered to me. He told me he had ordered Miss O'Neal to knock on Gowen's door at noon and to report to him her failure to get an answer. He sent her out to fetch the cop on the beat to be a witness, as one of his smaller bellboys scrambled through the transom of room 57 and discovered Gowen's body.

Wormley had the policeman remove the body at once, and take it to the precinct station. His staff then rolled up the bloody carpet and they were removing the decorative paper from the walls when Captain Linden's news reporter arrived. The newsman made sketches of the layout of the room and its furniture and interviewed Wormley and the hotel staff and rushed away to file his remarkable scoop. He reported that Franklin B. Gowen had shot himself behind locked doors. Mr. Wormley sent a telegram to the Gowen law firm to officially notify them of his death.

By the time the Coroner arrived at room 57, it no longer resembled the scene of the killing. The news report was already on the street and the pavement outside the hotel was crowded with the curious and the concerned. Some of Gowen's acquaintances had been permitted into the room and into the police station to view his body. Some voiced their opinion to the press that he was not the kind of man to commit suicide. Some insisted that he must have been murdered by the Molly Maguires. But they had no proof. The Coroner was convinced by Wormley that Gowen had taken his own life, and he certified to the same without doing an autopsy.

Later that evening, about eight o'clock, the Gowen nephews arrived by train from Philadelphia with Captain Linden. They informed the press that Linden had been placed in charge of the investigation. He announced that he would ferret out the Molly Maguires, if they had done the deed. All three came to the hotel. Linden interviewed Mr. Wormley and his staff, then the three continued on to the police station and to view the body. The nephews announced that since no autopsy was to be performed they would take the corpse of the former railroad President home by special train that very night. They left at 11:30 p.m. But Captain Linden did not go with them.

I had remained sequestered in my room all the time. At midnight Linden knocked on my door. Mr. Wormley had told me he would be coming. I let him in. Without offering any courtesies he thumped himself into my desk chair unbidden and I sat on the bed. He stared at me in silence for a long time. I stared back at him until, finally, he broke the silence.

"Mr. Wormley told my clients that their uncle had been shot by a street walker. When we were alone he told me that what happened was, the maid shot him. Depending upon . . . ah, the circumstances . . . that could be even more embarrassing to the family. He said that only you and the maid know why she shot him . . . that you saw the whole thing and I should talk with you."

"D'ye know who I am?" I asked him.

"I saw you at several of the trials. You made McKenna nervous, for some reason. He said you at one time accused him of being a detective but that he talked you out of the notion. I suspected there might be more he wasn't telling me."

"I suppose that's as close to the truth as McKenna ever gets," I said.

I went on to tell him about my manuscript, my encounters with Gowen; that the maid had shot him either accidentally or in self-defense while trying to get him to admit to his crimes.

He listened with rapt attention, interjecting only to help me clarify a point now and again.

When I was finished he asked, "Where's the manuscript?"

"It's safe," I advised him. "Why d'ye ask?"

He was exasperated by my question and seemed about to rebuke me, but he controlled himself somewhat and answered.

"Damnit, Mr. McWilliams! You're an Attorney. You must know that trying to make this look like a suicide is next to impossible. Gowen was shot above and behind the right ear, an impossible angle. And there are no powder burns at the point of entry. He was obviously shot by someone else who was at least an arms-length away."[3]

"I know," I said. "I learned all about powder burns in the Dan Dougherty case."

"Then enough said," Linden put up his hand with authority. "If I'm going to stick out my own neck and say this was a suicide, I need to be absolutely certain it wasn't a murder. I need to know if your bloody manuscript contains reasons for Gowen to strike out against

---

3 Editor's Note: For a penetrating analysis of the circumstances surrounding the shooting, see *Who Killed Franklin Gowen?*, by Patrick Campbell.

signing it. If not, I don't believe you!"

I went to the bureau and from under my clean shirts I pulled out the manuscript and handed it to Linden saying, "I'll be here through Monday. I have another copy. I expect you to return this to me by then."

"I'll be going to Philly at once and I'll be back by Monday," he said. "That will give me time enough to scan it."

My wait was not oppressive because there was no further need to keep to my room. I enjoyed the attractions of the Capitol city.

It wasn't until very late Monday evening that Linden returned with the manuscript and threw it onto the bed in my hotel room.

"You're right that Mr. Gowen never could have signed on to that," he admitted.

"He was infuriated by the prospect," I said.

Linden chortled and observed, "Mr. Pinkerton would have gone absolutely apoplectic over it. Would've been the death of him if he weren't already!"

"So what do we do?" I asked.

Linden sat himself in the chair, though I hadn't invited him, and he said, "Fortunately Mr. Wormley removed the body from the room and rolled up the carpeting and the wallpaper before the Coroner had a chance to examine the scene. After a few days of pretending to investigate, I'll announce that it was a suicide."

I walked to the door and opened it as an invitation for Linden to leave and said, "Good. Then we're done here."

"No, Attorney, we're not," he said and ordered, "Shut the damned door."

I shut it, went back to the bed to sit and asked, "What now?"

"The manuscript," he said. "The Agency doesn't want it published. If you don't agree to that we'll accuse the girl of murder, and perhaps yourself as well, and devil-take-the-hindmost. If you do agree, we want it in writing."

I was stunned for a long moment and then asked, "When d'ye need my answer?"

"Immediately," he replied.

I went to the door and again opened it and said, "I need to be alone for an hour."

He went. When he returned I had decided.

He sat again in the chair, again unbidden, and I resumed my seat on the bed.

"I agree upon two conditions," I said.

"Which are?" he demanded.

"First, the manuscript will be finished by me whenever I can get to it. Then it will be placed in a trust for fifty years along with a thousand dollars from the Agency to defray the cost of its publication when the trust is terminated."

"Not long enough," he shook his head emphatically. "Make it two hundred years."

"No," I refused without making a counter-offer.

"It must be put away for a longer time – at least until the twenty-first century," he insisted. "That is final."

I was disappointed, but then I considered that Miss O'Neal and her heirs might prefer it to be delayed that long.

"Agreed," I said.

"What else?" he asked.

"The Pinkerton Agency must purchase a well-appointed boarding house in Pottsville, Pennsylvania, and deed it over free and clear to Miss O'Neal and her daughter. That is final!"

Linden's face grew angry. But looking at me he saw that I was just as angry. He snarled, "Agreed! Be at the Agency's office in Philadelphia tomorrow."

He rose and strode out the door, slamming it behind him.

The day on which the Pinkertons and I signed our agreement, Captain Linden issued a press release to announce his formal finding that Franklin B. Gowen had committed suicide.

EPILOG

In 1890 the remnants of the Knights and the Amalgamated responded to the need for a single strong voice and organized the United Mine Workers Of America. The first local of the U.M.W. was formed in Shamokin in June of 1892, and, of course, my da' and Dutch Henry were there.

By the end of the 1890's Johnny d' Mitch had become President of the union. He fused its ethnic differences into solidarity with his rallying cry, "The coal ye dig isn't English or German or Italian or Polish or Irish coal, it's coal!"

There were setbacks like the Lattimer Massacre. But with the intervention of President Theodore Roosevelt on behalf of the miners in the strike of 1902, the legitimacy of a workers' union was finally established.

Miss O'Neal married a handsome Italian fellow who had lost his foot in the mines. They successfully operated their boarding house in Pottsville for many years. Her daughter, Molly, has long been married to a happy-go-lucky Irishman by the name of Maguire. Thus, it turned out that Gowen met his fate at the hand of the mother of Mrs. Molly Maguire. The irony of this occurs to me often. It sometimes vexes me that I can never mention it to anyone, but I've made my peace with that. The truth has now been written. I pray that future generations will divine from it whatever lessons they most require.

M. M.

THE END

NOTES TO THE TEXT

Notes To Chapter 7

Author's Note: 9

One of the most pitifully implausible of all the Allan Pinkerton misrepresentations is that the undercover detective whom he and Gowen had employed to infiltrate the Mollies had demanded and received from them, as a condition of his employment, their promise that he would never be required to testify in Court. (Pinkerton, pg. 504)

Pinkerton contradicts himself at page 18 of his book. On that page he states that he made Gowen consent that no Pinkerton detective should be required to testify unless Pinkerton should demand it. This arrangement would have been impossible for Pinkerton to abide by if Pinkerton had relinquished that discretion to the field operative.

The Gowen I knew would not have paid out good money for the services of an informer who would not testify in Court. Indeed, no reasonably competent businessman whose purpose was to convict the Mollies would have agreed to such an arrangement.

Finally, at page 330 of his book Pinkerton deviates into the truth. He recounts a conversation in which Pinkerton operative Captain Linden of the Coal & Iron Police warns the undercover McKenna about the reputation he has earned for himself:

"They – the police – look upon you as the worst and most desperate character in the Mollie crowd."

"I know it!" said McKenna, "but they'll learn their mistake one of these days!"

From this it appears that McKenna was in fact looking forward to the day he would reveal himself as a detective.

Author's Note: 10

Mr. Pinkerton's book (pp. 275-278) asserts that in the Agency's Philadelphia Office on April 18, 1875, he sent for Mr. Gowen who promptly arrived, and after Gowen left he summoned McKenna into his office.

The field report about McKenna written by Superintendent Franklin for April 28, 1875, summarizes the Pinkerton/Gowen meeting but says nothing about McKenna being present at the meeting.

I personally find it incredible that Pinkerton would cause his notorious but deep undercover agent to be traipsing around his office where the single glance of a knowing-eye could have betrayed his identity.

Superintendent Franklin's report to Gowen on operative McKenna for Sunday, May 2, 1875, when the meeting in Mount Airy among Pinkerton, Gowen and McKenna actually took place, is even more curious. It states:

> "This morning operative J. McF. met McAndrew at church, the latter told him he was going to Lanagan's Patch to see a man that was hurt from falling out of a buggy. The operative thinks there will be a great deal of trouble, as all the witnesses for the commonwealth committed perjury and knew they were doing so. Operative J. McF. does not think that all of Dougherty's witnesses swore to the truth."

The report that Mr. McAndrew had attended Mass and then went on a corporal work of mercy would have been of no interest to the busy Railroad President. I suspect it was inserted to assure Gowen that McAndrew was ready to swear that McKenna was in Shenandoah on the date of the meeting.

Editor's Note: 11

The sources which your Editor has been able to access contain many references to the men named in this sentence by the author. Coleman (pg. 160) states that the "Bum" was "one of the most depraved and infamous characters in the region" having been guilty of murder, highway robbery, larceny, assault and battery, mayhem, rape and attempting the subornation of perjury.

Tom Hurley had been a cohort of the Bum in Pottsville (Pinkerton, pp. 81, 82. 95) where they both met McKenna. Hurley had his own frequent business with the law and may be found in the newspapers as a cutthroat, literally. (*MJ* Oct. 10-15-75). For Bill Major we have cites already, as well as for Gomer James, the accused murderer of Cosgrove. "Bully" Bill Thomas is said by Broehl (pg. 213) to have earned his nickname with escapades of beatings, robbery and drunkeness. Pinkerton's book depicts Mike Doyle of Shenandoah as anxious to commit murder on various occasions. However, I was unable to find anything about this Mike Doyle in the public records of the time. This is not surprising since, as pointed out by Lavelle (pg. 276) the coal regions had many small newspapers, not all of which are now available.

Note To Chapter 12

Author's Note: 10

That Barney Dolan was a Pinkerton informer seems quite possible. As we have seen, he was responsible for installing both McKenna and Kerrigan into positions of influence within their local A.O.H. divisions. That McKenna continued seeking his advice, and that he was a violent man and an enemy of Kehoe's, is shown throughout McKenna's reports, especially in his report for July 1, 1875, and his expense account for July 4, 1875. The reports of Captain Linden reveal that when the policeman came to the anthracite fields, one of his first acts was to visit Barney Dolan for information. See: report of Linden for May 17, 1875, and May 21, 1875.

Note To Chapter 17

Author's Note: 4

WHO WAS THE LEADER OF THE SHENANDOAH GANG?

It seems clear from the trial testimony and the records heretofore cited that it was James McKenna, Secretary, and not Frank McAndrew, Bodymaster, who was *de facto* the boss of the Shenandoah Division of the A.O.H. That was certainly his reputation among Kehoe and Campbell and other A.O.H. members, some of whom, McKenna claims, confessed their worst crimes to him. Linden stated that his Coal & Iron Police considered McKenna, " . . . the worst and most desperate character in the Mollie crowd." (Pinkerton, pg. 330).

That he was actually the boss, as generally reputed, is further confirmed by Allan Pinkerton if one reads his book with both eyes open. At pages 201-203 he recites that Barney Dolan, then County Delegate of the A.O.H., asked McKenna to accept the Body-mastership. The detective refused the honor but then contrived to have it go to McAndrew who could neither read nor write and who would, therefore, be wholly dependant upon McKenna, who accepted the position of Secretary. It was a perfect position from which an agent provocateur could blame his crimes on someone else.

At page 146 Pinkerton affirms that McKenna was the most qualified man to hold the office of Bodymaster. At page 184 he actually calls McKenna "the Bodymaster". At 187-190 he depicts the men of the Shenandoah Division as being stricken with regret when McKenna left the town to visit other areas, imploring him to return and pressing upon him the gift of a blackjack. This is the send-off of a leader. At page 221 he recounts that the men of the Division, including Tom Hurley, did not think McAndrew fitted by nature or education to be Bodymaster.

Then there is the strange revelation at 216-217 of the book that on September 5, 1874, four new members were accepted into the Shenandoah Division but one man, a sibling of McAndrew's, was refused membership at McAndrew's insistence. He said his brother " . . .was continually in trouble and would surely bring disgrace upon the honorable brotherhood." This is amazing! McAndrew's brother was worse then Tom Hurley?! Worse than McAndrew himself who, according to McKenna, planned all the murders! The suspicious mind might infer that McAndrew, though himself hopelessly in the grasp of McKenna, refused to let the same thing happen to his brother.

Finally, there is the extraordinary testimony of McKenna that after Kehoe had denounced him to be a detective, he was fearful of being killed and McAndrew comforted him by saying:

"Have you got your pistols? Yes?! So have I and I will lose my life for you. I do not know whether you are a detective or not, but I do not know anything against you. I always knew you were doing right, and I will stand by you." (Comm. v. Kehoe Re: Thomas. pg. 95)

If this dialogue took place, it does not sound like a leader speaking to his underling, but rather it is blind sacrificial loyalty expressed to the speaker's acknowledged leader. Although it tends to prove my instant argument, the dialogue is somewhat preposterous. McAndrew was at least complicit in the many crimes of the Shenandoah Division. The only reason he would not have wanted McKenna dead and unable to testify would have been the existence of an agreement that the Pinkertons would not prosecute him. I believe this to be so, and the best proof of it is that McAndrew was not prosecuted. In this regard it should be noted that Division Secretary McKenna's minutes of what took place in Shenandoah have disappeared from the face of the earth. If he was an innocent detective, why did he not expose the minutes to the light of day? They would have been compelling evidence. The latter Division meetings were held in McAndrew's house; and McKenna testified that McAndrew took the Minute Book. (Comm. v. Kehoe Re: Bully Bill, pg. 72)

Did McAndrew hold McKenna's Minute Book as security for his freedom? In any event it seems clear that he was both stooge and accomplice to McKenna.

Note To Chapter 18

Author's Note: 7

To me, the chief significance of the trial of Kehoe and his Bodymasters for the assault upon Bully Bill is that it demonstrates McKenna's control over his gang and reveals that the real murderers were never tried because they were McKenna's accomplices. Whether Kehoe and/or his Bodymasters were guilty of ordering McKenna to make the assault is an issue secondary to my purpose.

However, full disclosure requires the notation that during the trial of the Bodymasters one of the accused men, Frank McHugh, was promised a light sentence for testifying on behalf of the prosecution and he did so. His testimony, like most testimony purchased by the Pinkertons, raised more questions than it resolved.

McHugh testified that he was Secretary of the Mahanoy City Division of the A.O.H. and a member for about eight months prior to the June 1 meeting. He had been brought in by Tom Hurley. His friend Hurley had also introduced him to McKenna. He hadn't been invited to the June 1 meeting but upon a chance encounter he was asked to get some writing paper and bring it into the room. When he brought it there one of the men, possibly McKenna, told him to take notes of the meeting. But the only events that he actually wrote in his minutes were that the meeting was called to order by Kehoe, and the date. So his minutes do not corroborate his testimony as to what happened.

He said that Dan Dougherty was sent for and Dan named Jesse Major as the one who shot at him. This leaves it a mystery why Bully Bill should then be selected as someone to be killed in retaliation.

He said that after Dougherty left the meeting it was decided that the Major brothers and Bully Bill were to be shot. This was because of their general conduct as dangerous men. McKenna was put in charge of killing Bully Bill. Other men were to get the Major brothers.

He claims that those murder plans were the only things discussed at that meeting. He spent most of the time during the meeting in a private conversation with the County Delegate of Northumberland County about how things were faring there.

He said he was asked to write down minutes of the meeting, "To make it appear lawful." He left his writing at the meeting, after putting down the names of those who were there.

On cross-examination he stated that during the entire term of his membership in the A.O.H., the June 1 meeting was the only occasion when he heard the organization encourage any crime. He admitted that he expected less punishment than if he had not testified for the prosecution. No assault was ever perpetrated against the Major brothers.

The Pinkerton/Gowen version of this June 1 meeting makes it a most unusual gathering.

None of the details of the workingmen's solidarity parades or the picnics, or the dances which the men participated in over the next few days were discussed at all. The men allegedly came from all over the coal region merely to watch Dan Dougherty show them bullet holes in his coat put there by Jesse Major and to listen to Kehoe order the detective to go out and immediately kill a different workingman, Bully Bill. If a three-County meeting was actually necessary to authorize this murder, then were all the other murders un-authorized?

The Pinkertons want us to believe that the Bodymasters assembled in order to plan cold-blooded murder and they were so frightful of getting caught that they took a very stupid precaution. They went out and buttonholed a young fellow who had not been invited to the meeting to take down all their names and falsify minutes to make them appear lawful. Yet they didn't bother to confront him when he flatly failed to make the false entries. Was that because he was such a good friend of McKenna's good friend, Tom Hurley? Or did McKenna order him to pretend to take minutes because McKenna calculated he could later force him to say whatever was necessary for the Pinkerton agenda?

Finally to be noted is one last remarkable fact. One of the Bodymasters who had been summoned to attend was Powderkeg Kerrigan. And Kerrigan did respond. He traveled to Mahanoy City all the way from Tamaqua. But upon his arrival, Pinkerton claims, he stayed in Clark's barroom all the time and never went upstairs to the meeting. (Pinkerton, pg. 307). So, of course, he didn't have to be tried and thrown in jail with the rest of them. What a wonderful series of bizarre actions and coincidences fell in favor of the Pinkertons and their cohorts on that remarkable day!

APPENDIX A (i)

Pinkerton's operative reports for McParlan – July 1, through July 4, 1875.

PHILA. JULY 31ST. 1875.

F. B. GOWEN ESQ.

PREST. PHILA. & R.R.R.CO.

PHILA. PA.

DEAR SIR

ANNEXED PLEASE FIND A SYNOPSIS OF THE REPORTS OF OPERATIVE J. MC, F. WHO HAS BEEN DETAILED ACCORDING TO YOUR ORDERS, TO VISIT THE COAL REGIONS OF PENNSYLVANIA, AND ASSOCIATE WITH THE "MOLLY MAGUIRES" AND OBTAIN INFORMATION REGARDING THEM.

THE OPERATIVE IS AT PRESENT LOCATED AT SHENANDOAH PA. AND IS A MEMBER OF A BODY OF "MOLLY MAGUIRES" AT THAT PLACE OF WHICH FRANK MC, ANDREWS IS THE CHIEF.

THE PRESENT REPORT CONTAINS A SYNOPSIS OF THE INFORMATION GAINED BY THE OPERATIVE, FROM JULY 1ST. TO JULY 24TH. 1875. INCLUSIVE.

RESPECTFULLY SUBMITTED.

ALLAN PINKERTON.

PER BENJ. FRANKLIN.

SUPT.

REPORT OF J.MC,F

THURSDAY JULY 1ST. 1875.

BARNEY DOLAN WAS IN SHENANDOAH TO-DAY BUYING UP CITY WARRANTS. HE HAS ABOUT $8000, IN BONDS AND WARRANTS IN THE BOROUGH.

HE TOLD THE OPERATIVE THAT THE AFFAIR AT MAHONEY CITY WAS A FAILURE. HE SAID HE HAD A CROWD OF ABOUT TWENTY-FIVE MEN, WHO ALWAYS DID SUCH JOBS FOR HIM, BUT SAID KEHOE COULD NEVER DO ANYTHING HIMSELF AS HE WAS TOO COWARDLY. HE REPORTED ALL QUIET IN HIS SECTION, AND SAID THEY WERE WORKING AT THE BEAR.

THE OPERATIVE ASSISTED MC,ANDREWS TO-DAY IN MAKING UP THE MOLLY MAGUIRE BOOKS, BUT LEARNED NOTHING NEW FROM THEM. MC,ANDREWS HAS JUST RENTED A HOUSE ON CENTRE ST. WHICH CONTAINS A LARGE ATTIC, AND THEY INTEND TO USE IT AS A PLACE OF MEETING FOR THE "MOLLY MAGUIRES".

THE FRESYTHE AFFAIR IS POSTPONED UNTIL THE NIGHT OF FRIDAY THE 9TH. OF JULY, WHEN THEY INTEND TO ATTACK HIM AS HE COMES FROM LODGE.

HURLEY WORKS AT THE KOHINOOR, AND LIVES AT THE WEST END OF CENTRE ST. NORRIS WORKS AT THE PLANK RIDGE, AND BOARDS AT A SALOON KEPT BY ONE PATSEY MACK, NEAR THE PHILA'S READING TERMINAL. DOYLE BOARDS AND ROOMS WITH THE OPERATIVE.

FRIDAY JULY 2ND. 1875.

TO-DAY OPERATIVE J.MC,F HELPED MC,ANDREWS ABOUT MOVING, AND SPENT THE EVENING IN LOAFING AROUND WITH DOYLE, HURLEY, AND SEVERAL OTHERS BUT STILL HAS NOTHING NEW TO REPORT.

SATURDAY JULY 3RD. 1875.

TO-DAY MC,ANDREWS INVITED THE OPERATIVE TO GO WITH HIM TO-MORROW TO GIRARDVILLE AND BIG MINE RUN. THEY ARE TO GO IN THE MORNING. MC,ANDREWS SAID HIS OBJECT IN GOING TO THE RUN WAS TO GET SOME THREE OR FOUR MEN, TO KEEP WATCH OF GOMER JAMES, AT THE PICNIC TO BE HELD MONDAY AT NO.3. HILL AND IF A GOOD CHANCE OFFERED ITSELF, TO FINISH HIM.

THIS EVENING ED.FERGUSEN, WHO IS WORKING AT ASHLAND, WAS IN TOWN. THE "MOLLY MAGUIRES" OF SHENANDOAH WERE ALL OUT. A LARGE NUMBER OF THEM MET IN CUFFS SALOON, WHERE THEY WENT FOR DRINKS. MC,ANDREWS TOLD DOYLE THAT HE WAS A COWARD, AND DID NOT HALF DO HIS WORK IN THE THOMAS CASE.

CONSIDERABLE DISSATISFACTION WAS MANIFESTED BY THE "MOLLY MAGUIRES" IN REGARD TO THE STATE OF THEIR AFFAIRS AT THE PRESENT TIME. IT IS EVIDENT THAT THINGS ARE NOT EXACTLY AS THEY WOULD LIKE TO HAVE THEM.

SUNDAY JULY 4TH. 1875.

THIS MORNING OPERATIVE J.MC,F. MC,ANDREWS AND MORRIS WENT TO GIRARDVILLE, THERE THEY SAW DONOHOE, PHIL,NASH, AND A GREAT MANY OTHERS OF THE "MOLLY MAGUIRES". DONOHOE TOLD THE OPERATIVE THAT DOUGHERTY WAS AT LOCUST GAP. HE SAID THAT KEHOE WAS GOTTEN HIM 'DONOHOE' A JOB, AND HE SHOULD STICK TO IT, SAID HE WAS VERY POOR AND COULD NOT LIVE LONGER WITHOUT WORK. HE HAS NOT BEEN AT LOCUST GAP SINCE THE OPERATIVE LAST SAW HIM THERE. HE SAID HE SAW GIBBONS WHEN HE WAS GOING AWAY, THAT KEHOE GAVE HIM $1.50, AND HE DONOHOE GAVE HIM $2.00 AND GOT A HORSE & CARRIAGE AND TOOK HIM TO RUPERT STATION, WHERE HE TOOK THE TRAIN FOR WILKESBARRE, AND IS THERE AT PRESENT.

THEN MC,ANDREWS AND THE OPERATIVE INFORMED KEHOE WHAT WAS THE OBJECT OF THEIR VISIT TO GIRARDVILLE, HE SAID THEY COULD NOT GET A MAN THERE FOR THE JOB WHO WAS WORTH ANY THING, BUT THOUGHT PERHAPS THEY COULD AT BIG MINE RUN. THEY NEXT VISITED HOMERVILLE AND CALL AT MORRIS CONWAYS. DOLAN AND A GREAT MANY OTHER "MOLLY MAGUIRES" WERE THERE WITH WHOM THEY HAD A TALK. THEY WANTED TO SEE PAT DOLAN, THE BODY MASTER BUT HE WAS NOT AT HOME, SO THEY RETURNED TO KEHOES, AND THERE SAW LARRY CREAN, BODY MASTER OF GIRARDVILLE. MC,ANDREWS TRIED TO OBTAIN MEN OF HIM, FOR THE ABOVE NAMED PURPOSE, BUT HE REFUSED TO LET HIM HAVE ANY UPON WHICH MC,A. BECAME VERY ANGRY. HE TOLD CREAN THAT HE KEPT TRAITORS AROUND HIM, AND THAT IT WAS TWO OF HIS MEN WHO BETRAYED KEHOE, AND CAUSED FATHER BRIDGEMAN TO PUBLISH HIM AND CURRAN FROM THE ALTAR. AND MC,ANDREWS FARTHER SAID IF HE HAD SUCH MEN AS THAT HE WOULD SHOOT THEM.

ABOUT THIS TIME WILLIAM GAVIN COUNTY SECRETARY, AND PAT.DOLAN CAME IN, AND DOLAN SAID HE WOULD SEND TWO MEN, AND ALTHOUGH AN OLD MAN FOR SUCH WORK WOULD MAKE THE THIRD HIMSELF.

ON THEIR RETURN TO SHENANDOAH, HURLEY AND BUTLER WERE OF THE PARTY. BUTLER SAID ON FRIDAY NIGHT HE WOULD TAKE FIVE MEN WITH HIM AND THEY WOULD FINISH FRESYTHE, AT ALL HAZARDS.

HURLEY SAID HE THOUGHT THEY WERE TOO FAST IN THIS MATTER, HE SAID F. DISCHARGED GARVEY FOR INSOLENCE, AND IF HE HAD BEEN BOSS HE WOULD HAVE DONE THE SAME. AND THAT FRESYTHE HAD ALWAYS ACTED SQUARELY WITH HIM AND ALL THE REST OF THE MEN. BUTLER SAID IF THAT WAS THE CASE HE WOULD NOT INTERFERE, BUT THEY COULD TALK THE MATTER OVER, AND HE WOULD SEE THEM ALL TO-MORROW NIGHT, AT THE DANCE AT NO.3 SCHOOL HOUSE, AND WOULD BE SATISFIED WITH ANY PLAN WHICH THEY AGREED UPON.

APPENDIX A (ii)

McKenna's Expense Account for July 1875

Pinkerton's National Detective Agency

ALLAN PINKERTON, *Principal*
GEO. H. BANGS, *Gen'l Sup't.*

CHICAGO:	NEW YORK:	PHILADELPHIA:
191 & 193 Fifth Ave.,	66 Exchange Place,	45 S. Third Street,
F. WARNER, Sup't.	R. A. PINKERTON, Sup't.	BENJ. FRANKLIN, Sup't.

CLARENCE A. SEWARD, Attorney and Counsel for the Agency, 59 Nassau Street, New York.

To Services J.M.G. July 1 Am To Sept 30th 1875. 92 days @ $6		552	00	
Expenses of same	Item 1	298	64	
" Telegraphing	Item 2	2	08	
				$852 72

		Item 1	J.M.G.	
1875				
June	29	Treating Thompson & five friends at Cleary's	55	
		" Mike Doyle & two friends at Riley's	35	
	21	" " " friends	30	
		" Pat Clark & friend at Riley's	95	
July	1	" Pat Butler and 6 friends at Johns	65	
		" Frank McAndrew & friend	45	
	2	Washing	75	
		Treating Mike Doyle & friend at Cleary's	45	
	3	" Ed Mooney & friend	45	
		" Crowd of M.M's at Cripps's	95	
	4	Share of buggy hire to Big Mine Run	2 00	
		Treating Donahue & crowd of M.M's	1 25	
		" Frank McHugh & crowd of M.M's	85	
		" Crowd of M.M's at Big Mine Run	1 50	
	5	1 Weeks Board & Room	7 00	
		Treating McAndrew & friend	60	
		" Canoll and friend	45	
	6	Treating Frank McAndrew & friend	65	
		" Jno Harris "	55	

Pinkerton's National Detective Agency

ALLAN PINKERTON,
PRINCIPAL.
GEO H BANGS, Gen'l Sup't.

CHICAGO: NEW YORK: PHILADELPHIA:
191 & 193 Fifth Ave.,   66 Exchange Place,   45 S. Third Street,
F. WARNER, Sup't.   R. A. PINKERTON, Sup't.   BENJ. FRANKLIN, Sup't.

CLARENCE A. SEWARD, Attorney and Counsel for the Agency, 29 Nassau Street, New York.

Philadelphia Oct 11th 1875

Phila & Reading C. & I. Co.

Allan Pinkerton,

1875

For Services and Expenses incurred
in making investigation concerning
the secret organization known as
the Molly Maguires, in Schuylkill
Co. Penna.

To Services J McPd July 1 Am To Sept 30th th. 92 days @ $6		552	00
" Expenses of same	Item 1	298	64
" Telegraphing	Item 2	2	08
			$852 72

Rec'd Payment

EDITOR'S LIST OF CITATIONS AND ABBREVIATIONS

NEWSPAPERS
Pottsville Daily Miners' Journal (MJ)
Pottsville Evening Chronicle
The Carbon Advocate
Mauch Chunk Gazette
Mauch Chunk Democrat
Shamokin Citizen
Shenandoah Herald
New York Times
New York Tribune
Philadelphia Evening Express
Philadelphia Inquirer

TRIAL TRANSCRIPTS OTHER THAN NEWSPAPERS

Commonwealth v. Alexander Campbell Re: John P. Jones	Carbon County Clerk of Courts
Commonwealth v. Alexander Campbell Re: Morgan Powell	Carbon County Clerk of Courts
Commonwealth v. Michael J. Doyle Re: John P. Jones	Carbon County Clerk of Courts
Commonwealth v. John Kehoe Re: Wm. M. Thomas	Pottsville: Miners' Journal Book & Job Rooms, 1876

DETECTIVE REPORTS & EXPENSE ACCOUNTS
Hagley Museum and Library, Wilmington, Delaware:
Philadelphia & Reading Railroad Collection
Historical Information
Molly Maguire Papers

See also: Historical Society of Schuylkill County

BOOKS

Aurand, Harold W.	From The Molly Maguires To The United Mine Workers
Barrett, Tom	The Mollies Were Men
Bimba, Anthony	The Molly Maguires
Broehl, Wayne G.	The Molly Maguires
Campbell, Patrick	A Molly Maguire Story Who Killed Franklin Gowen?
Coleman, James W.	The Molly Maguire Riots
Crown, H. T. with Major, M.T.	A Molly Maguire On Trial, A Guide To The Molly Maguires
Dewees, Francis P.	The Molly Maguires
Hitler, Adolph	Mein Kampf
Kenny, Kevin	Making Sense Of The Molly Maguires
Korson, George G.	Minstrels Of The Mine Patch
LaVelle, Hon. John P.	The Hard Coal Docket
Lewis, Arthur H.	Lament For The Molly Maguires
MacManus, Seamus	History Of The Irish Race
Miller & Sharples	The Kingdom of Coal
Pinkerton, Allan	The Molly Maguires and the Detectives
Wallace, Anthony F. C.	St. Clair

INDEX

A

Amalgamated Miners' Assn., The, 218, 241
Ancient Order of Hibernians, iii, 46-49, 56, 57, 125, 130, 133, 134, 148, 152, 157, 170, 194, 203, 206, 210
Anthracite Coal, v, 6
Anthracite Mines and Mining,
        business conditions, v-vi,
        life expectancy of miners, vi,
        employment in, 5-7,
        working in, 19-23,
Anti-Catholic bias, 157, 162, 182, 205
Ashland, 7, 46, 66, 83, 88, 94, 96

B

Bannan, Benjamin, iii
Bartholomew, Lin, 170-172, 174, 175, 178, 182
Big Mine Run, III, 140, 141
Boyle, James, 191
Broehl, W. G., III
Burke, Edmund, 212

C

Calhoun, John C., 13
Campbell, Alex, III, i, 133, 154, 209,
        his implication in the future murder of Jones, 144, 145,
        arrested for Jones murder, 172,
        implicated by the press prior to trial, 173-175, 182,
        accused by McKenna of complicity in Yost murder, 177-179,
        implicated by Kerrigan in Yost case, 181-182,
        evidence exonerating him in Jones' murder, 183-184,
        jury prejudice in Jones case, 185-187,
        his trial for the Jones murder, 189-205,
        his hanging and lasting impression, 211, 212
Campbell, Patrick, VI
Carbon County, 132, 163, 181, 212
Carroll, James, 130, 131
Centralia, 7, 30, 68, 85

Civil War:
        draft resistance, iii,
        Gowen opts out of, 17,
        Gowen's brother killed in, 18
Clark, Michael, 29, 30, 47-51
Coaldale, 109, 129
Coal Township, 67
Collieries (Coal Breakers);
        incinerated, iv, 82, 162
Cosgrove, Edward, 59, 146, 147

Darrow, Clarence, 213
Denver, Colorado, 213
Dolan, Barney, 133
Dolan, "Bear", 46
Dougherty, Daniel, II,
        the Major murder, 31-45, 48-53, 55, 56, 57, 120, 170, 191, 238,
        trial of, 57-64,
        declaration of his innocence, 107, 108
Doyle, Michael, J., i, 157, 199, 200
Doyle, "Mickey", v, 78, 115, 117, 122, 149, 150, 151, 162
Doyle, Sir Arthur Canon, I

Evening Chronicle, The, 147, 148, 173, 176, 179

Franklin, Benjamin, 195, 196, 198, 201

Gibbons, John, 122
Girardville, III, 46, 88, 96, 97, 139, 140, 148, 158
Gowen, Franklin B., I, V, VI, ii, iii, iv, v, vi, 7, 19, 56, 64-65, 132, 141, 142,
        145, 152, 170, 203-208, 211, 212-215, 217-224,
        early years in Shamokin, 9-16,
        in Pottsville, 17-18,
        the W.B.A., 24-25,
        home in Mt. Airy, 68-72,
        speculation as to his demands on the Pinkertons, 72-76,
        the Wm. Thomas assault trial, 111-113,
        speculation re: his death, 251-241

H

Hazleton, 66, 105
Hester, Patrick, i, 210, 211
Holt, Mrs. E., 36, 62
Hughes, Francis W., 58-60, 63, 64, 191
Hurley, Thomas, v, 49, 53, 56, 78, 115-118, 122, 147-151, 205

I

Immigration; Irish, 1

J

James, Gomer, 59, 78, 96, 139-141,
        his murder, 147-148
Jones, John P., 132, 137, 139-141, 162, 174, 178,
        his murder, 142-146, 152-156,
        trial of Alex Campbell for his murder, 189-204

K

Kehoe, John, III, i, ii, 46-53, 57, 125, 133, 147, 148, 159-164, 166, 169, 176,
        his support of the strike, 109, 110,
        his trial for the assault on Wm. Thomas, 110, 111,
        McKenna's false reports concerning, 139-141,
        his trial and conviction for an 1862 murder, 208-210
Kelly, Edward, i, 157, 199, 200
Kelly, the "Bum", 47, 57, 78, 210, 211, 213
Kerrigan, James, 110, 129, 134, 142, 162, 172-175, 176-178, 182-185, 208,
        his murder of Officer Yost, 124-127, 130-133,
        confrontation in the Tamaqua Cemetery, 134-139,
        his murder of Jones and his arrest, 145, 152-157,
        his testimony in the Yost case, 175, 181,
        testimony about him in Com. v. Campbell re: Jones, 189, 193-203
Knights of Labor, The, 217, 218, 241
Koch, Rev. Joseph, 47-48, 210

L

Lansford, 7, 130, 152-154, 202
Latimer, 241
Linden, Robert J., III, IV, V, 83, 85, 94-96, 114, 120, 123, 134-139, 142, 145,
        153, 155, 195-199, 201, 202, 208, 223, 233, 234, 237-240
Locust Gap, 7, 46, 66, 68, 83, 88, 109, 146, 162, 210
Long Strike, 24, 65-68, 81-86, 105, 106
Lost Creek, 47, 57
L'Velle, Martin, 207

M

Mahanoy City, 7, 25, 57, 64, 78, 109, 116, 121, 148, 169, 195,
     the Major murder, 27-45
Major, George, 29, 51, 53, 55, 56, 59,
     murder of, 34-44
Major, Jesse, 29, 30, 100
Major, William, 30, 34, 35, 37-40, 42, 45, 50, 58, 60, 61-63
Mauch Chunk, 203, 211
McAndrew, Frank, 115, 122, 153, 195
McCann, John,
     the Major murder, 31-43, 45, 50, 51, 56, 58, 60-61, 64
McCarron, Barney, 125-127
McGeehan, Hugh, 132, 133, 144, 179, 184, 189, 191-193
McKenna, James, IV, 49-54, 57, 78, 93-96, 105, 124, 125, 161, 162, 170,
     205-208, 210, 211, 222, 238,
        speculation as to his early discovery, 72-78, 97-103,
        speculation as to how he was tricked into making false reports, 88-94,
        resulting meeting with Capt. Linden, 95, 96,
        conspiracy to murder Wm. Thomas, 109-122,
        his testimony re: the A.O.H., 111-114,
        his leading of the assault on Thomas, 114-120,
        his investigation of the Yost murder, 129, 130,
        his plot to murder Gomer James, 139, 140,
        his reports, 139-141,
        his attempt to implicate Campbell in a future murder, 144-145,
        the murder of Gomer James, 146-148,
        the Raven Run murders, 149-151,
        the murder of John P. Jones, 152, 153,
        his true identity made public, he flees, 174,
        his testimony in the Yost case, mostly against non-defendants, 175-179,
        his testimony in Com. v. Campbell re: Jones, 190, 203
McParlan, James, a/k/a McKenna, James, III, iii, v, 213
Miners' Journal, II, ii, vi, 56, 64, 66, 67, 116, 120, 166, 169, 170, 173
Modocs, 28, 51, 55, 59, 64, 78, 110, 134, 201
Morris, John, 122
Mt. Airy, 68
Mt. Carmel, 7, 67
Mt. Laffee, 109, 155, 199
Munley, Thomas, 205

O

O'Donnell, Charles, murdered, 163-168
O'Donnell, Ellen, murdered, 167

O'Donnell, "Friday", 163-168
O'Donnell, the clan, 134, 149, 150, 170, 172,
        targeted by confidential memo, 163-165,
        murder of its members, 166-168

Philadelphia & Reading Railroad, III, 18, 24, 27, 214, 217
Pinkerton, Allan, III, ii, 133, 213, 222
Pinkerton's National Detective Agency, IV, ii, 64, 120, 121, 127, 132-134,
        143, 144, 151, 153-157, 162, 166, 189, 202, 203, 205-207, 211, 217
Port Carbon, 121
Pottsville, III, 56, 57, 155, 211, 241,
        Gowen's early years in, 17, 18

Raven Run, 149, 150, 151, 152, 163-168, 170, 173, 178, 202, 205, 206
Roarity, James, 129, 130, 177
Roosevelt, Theodore, 241

Sanger, Thomas, murder of, 149-151
Schuylkill County, iv, 56, 65, 105, 132, 193, 199, 207
Scranton, 217
Scranton, William, 217
Shamokin, III, iv, v, vi, 65, 67, 217, 221, 227, 241,
        Gowen's early years in, 9-17
Shenandoah, 7, 51, 57, 78, 123, 132, 146, 149-151, 153, 194, 196, 201
Shenandoah Herald, ii, 56, 166, 173
slavery, debate over,
        Gowen's rhetoric in support of, 9, 10, 13,
        Gowen's true position re: 11, 12, 17
St. Clair, 109, 115, 221
Summit Hill, 7, 190
Swampoodle slum, D.C., 234, 235

Tamaqua, 7, 14, 125, 129, 141, 145, 153, 155, 181, 192, 193, 195, 196, 199
Thomas, Wm. "Bully Bill", 78, 124, 148, 149, 178, 206, 223,
        assault upon, 109, 110, 114-122
Trevorton, 7

United Mine Workers of America, 241

Washington, D.C., 218
Wiggans' Patch, 134, 149-151, 163-168
Wood, Archbishop, 170
Workers Benevolent Association, 24, 28, 48, 50, 65, 66, 70, 71, 87, 105,
        123, 164,
            effect of its demise, 123-124,
            its replacement by other unions, 217, 218, 241
Wormley, James, 218, 219,
            the death of Gowen, 229-234, 236, 237, 239
Wormley Hotel, V, 218, 221, 223, 233

Yost, Frank, 125, 144, 154, 172, 191-193,
            his murder, 125-127,
            the investigation of his murder, 129-134,
            the trial re: his murder, 177-179